No. 1005
$9.95

THE HANDBOOK OF
SOLAR FLARE
MONITORING & PROPAGATION
FORECASTING

By Carl M. Chernan

TAB BOOKS
Blue Ridge Summit, Pa. 17214

FIRST EDITION

FIRST PRINTING— MARCH 1978

Library of Congress Cataloging in Publication Data

Chernan, Carl M.
  The handbook of solar flare monitoring & propagation fore-casting.

  Includes index.
  1. Solar flares—Observers' manuals.
2. Radio—Apparatus and supplies. I. Title.
QB526.F6C48   523.7'5   77-19026
ISBN 0-8306-9984-8
ISBN 0-8306-1005-7 pbk.

Solar prominence photo by NASA. Print material supplied by courtesy of Matthey Bishop, Inc.

DEDICATION

This book is dedicated to Professor Jaroslav J. Slezak of St. Vincent College (Physics Department), Latrobe, Pennsylvania. Without his help and encouragement this book would not have been possible.

Preface

Many serious radio amateurs and experimenters who become interested in solar flares and their effects on radio propagation usually are stymied by the lack of instruments for observing this most fascinating phenomenon of the sun.

Commercial instrumentation for solar flare detection is complex and costly; however, flares can be detected by their effect on the earth's ionosphere using remarkably simple radio equipment. Although these indirect methods of detection lack the obvious advantage of actually seeing the flare occur, they do have some other advantages in addition to being simple and economical. One very important advantage is that they can record the occurrence of solar flares automatically and can run unattended for several days, rather than requiring the constant attention of the observer. The purpose of this book is to acquaint the amateur and the experimenter with the SEA and SES methods of flare detection, including the construction of three solid state VLF receivers and recording interfaces. Welcome to the serious, but light, study of the following pages, bringing together an elementary knowledge of radio propagation, antennas, commercial receiving gear, and an introduction to the "sub-basement" radio frequencies. Hopefully this will lead to many pleasurable hours of enjoyment in operating a solar flare recording station.

Carl M. Chernan WA3UER

Contents

Chapter 1

Below The Broadcast Band

Anyone who has listened to a short wave receiver is well aware of the vagaries of high frequency radio transmissions. While average propagation conditions can be predicted with reasonable certainty, high frequency circuits are still plagued by erratic signals and sudden blackouts. But there are frequencies where this is not true.

DOWN UNDER

The glamour and fascination of the satellites and the space age have tended to overshadow the less known but vitally important very low frequency (VLF) services which have offered dependable, world wide communications for over forty years. This spectrum is now packed with a variety of CW and RTTY (very narrow shift) signals. Because of the propagation characteristics of the very low frequencies, many of the super power nations use powerful transmitters operating below 30 kHz to communicate with their patrolling nuclear submarines. Transmitters with input powers of 1,000,000 watts are not uncommon. Below 14 kHz there are highly sophisticated navigational systems (Omega for the United States) operating around the world. Radio location and navigational systems also use frequencies around 70, 85, and 100 kHz.

The range of 200–400 kHz is used for aircraft low frequency direction finding services. These low power transmitters can be heard within a radius of 100 miles daytime and 1000 miles or more at night. Detailed aviation weather forecasts are broadcast in voice from the larger cities. Frequencies around 500 kHz are used for marine traffic, with 500 kHz itself reserved as a calling and distress frequency. There are many radio (both SWL and amateur) old timers who vividly remember the long wave radio stations in their day, which used to provide hours of listening enjoyment in the form of code practice material, regular press copy, weather information, and time checks. There were also many newcomers to radio who became familiar with the low frequency services (with the multitude of marine radio beacon, aeronautical radio beacon, and aeronautical range stations) through their work in the armed forces during World War II and the Korean conflict. Even so this author has found that there are many persons actively engaged in electronics who are completely unaware of the considerable activity "below the broadcast band." This chapter will give you an idea what goes on "below the broadcast band" and show you how some of the frequencies can be put to use monitoring solar flares.

FREQUENCIES AND THEIR USES

First of all, what are these low frequencies, and why would anyone want to operate a transmitter down there? The nomenclature of these bands is covered in Table 1-1.

The uses to which the frequencies from 10 to 510 kHz may be put are determined by international treaty. These allocations are shown in Table 1-2. The frequencies from 30 to 300 kHz are normally used for long distance ground wave communication. Frequencies in the medium band, covering 300 to 3000 kHz, are usually used for long distance communication over sea water or medium distance communication over land. Use of the ground wave for transmissions at low frequencies, providing sufficient power is used, permits 24 hour a day coverage of large global areas, whereas sky

Table 1-1. Nomenclature of Low and Very Low Frequency Bands.

Frequency Range	Frequency Subdivision
3—30 kHz	Very low
30—300 kHz	Low
300—3000 kHz	Medium

Table 1-2. Partial Listing of Various Services and Their Frequency Allocations.

kHz	SERVICE
10—14	Radio navigation
14—90	Fixed, maritime mobile
90—110	Fixed, maritime mobile, radio navigation
110—160	Fixed, maritime mobile
160—200	Fixed
200—285	Aeronautical mobile, aeronautical navigation
285—325	Maritime navigation (radio beacons)
325—405	Aeronautical mobile, aeronautical navigation
405—415	Aeronautical mobile, aeronautical navigation, maritime navigation (radio direction finding)
415—490	Maritime mobile
490—510	Mobile (distress and calling)

waves cause fading and are subject to daily, seasonal, and other variations due to changes of ionospheric conditions.

Many years ago, communicators found that the use of frequencies well below 500 kHz in the northern latitudes (above 60° North) was an effective means of maintaining radio communications during the magnetic disturbances or auroral conditions which at times prevented propagation of the higher frequencies. Since very low frequencies make use of the ground wave for transmission, stations operating on these frequencies may be found coming through with fine signals morning, noon, and night all through the year.

As far back as 1923 AT&T was operating a single sideband transatlantic radio telephone circuit on 55 kHz. Even after the transatlantic and other circuits shifted operations to higher frequencies, the low frequency station was maintained as an emergency link with Europe when magnetic storms would virtually wipe out all higher frequency communications.

Not long ago, in November 1960, high frequency radio conditions were the poorest that they had been in six years. A solar eruption took place on November 12 and on November 13 a high frequency radio blackout existed in almost all areas of the world. During this time VLF frequencies were used without difficulty. The reason for maintaining these very low frequency stations is obvious, especially today when breakdown of foreign communication circuits is intolerable. There are, however, several other reasons for maintaining the existing stations and also adding new and higher powered stations to the VLF spectrum.

The U.S. Navy operates quite a number of stations in the very-low and low frequency bands. These stations are capable of running tremendous power in order to transmit information to Navy submarines lurking some 90 feet or better below the ocean surface. One such station is located at Jim Creek, Washington, and can deliver a megawatt into its antenna system. This station is presently being keyed via high frequency radio link from San Francisco and uses the call NPG. Another Navy station NBA in Panama transmits timing signals on 18 kHz. These signals consist of a keyed carrier of 300 milliseconds duration, conforming to a particular pattern. This station operates 24 hours a day, seven days a week. WWVL, the National Bureau of Standards station at Boulder, Colorado, has been transmitting on 20 kHz since April 5, 1960. WWVL transmissions have been heard as far as 9000 miles from Boulder.

WWVB, another Bureau of Standards station at Boulder (formerly KK2XEI), has been operating on 60 kHz with an effective radiated power of 3 kW. This station can be received at distances up to 1700 miles, using specialized narrow band receivers.

One might wonder why these standard-frequency stations are operating at such low frequencies. The reason is that the 20 kHz and 60 kHz transmissions get away from the small, but significant, errors that result from ionospheric propagation of the higher standard-frequency transmissions. The need for extremely high accuracy in frequency and time measurements, together with the necessity of covering the globe irrespective of propagation conditions, is the prime reason for the shift to VLF. In addition to providing the military with useful communication channels, the lower frequencies supply much useful material for wave propagation studies.

The experimenter can make good use of monitoring 18.6 kHz to study solar flare propagation. The ham or would-be ham who wishes to get in some code copying practice can find any number of civilian or military stations on the air sending excellent practice material in the form of regular press copy or code groups. Again for the ham or experimenter, there are many stations sending RTTY signals. These stations are on continuously, making it possible to copy for extended periods of time in order to make adjustments to one's equipment. The graph shown in Fig. 1-1 shows the complete LF and VLF spectrum from 0 to 600 kHz, listing the stations and their approximate frequencies. A more accurate and detailed description of LF and VLF stations showing call, frequency, location, and transmitting power can be found in Appendix A.

WAVE PROPAGATION

In order to understand the method of detection of solar flares using low frequencies as a source of indirect monitoring, a brief summary of wave propagation must be understood.

Characteristics of Radio Waves

Radio waves, like other forms of electromagnetic radiation (such as light), travel at a speed of 300,000 meters per second in free space and can be reflected, refracted, and also diffracted.

An electromagnetic wave is composed of moving fields of electric and magnetic force. The lines of force in the electric and magnetic fields are at right angles and are mutually perpendicular to the direction of travel. A simple representation of a wave is shown in Fig. 1-2.

In this drawing the electric lines are perpendicular to the earth and the magnetic lines are horizontal. They could, however, have any position with respect to the earth as long as they remain perpendicular to each other. The plane containing the continuous lines of electric and magnetic force shown by the grid or mesh-like drawing in the figure is called the wave *front*. The mediums in which electromagnetic waves travel have a marked influence on the speed with which they move. When the medium is empty, the speed is 300,000 meters per second. It is almost, but not quite, that great in air and is much less in some other substances. When a wave meets a good conductor, it cannot penetrate it to any extent (although it will travel through a dielectric with ease) because the electric lines of force are practically short circuited.

Polarization

The polarization of a radio wave is taken as the direction of the lines of force in the electric field. If the electric lines are perpendicular to the earth, the wave is said to be vertically polarized; if parallel with the earth, the wave is horizontally polarized. The longer waves, when traveling along the ground, usually maintain their polarization in the same plane as was generated at the antenna. The polarization of shorter waves may be altered during travel, however, and sometimes will vary quite rapidly.

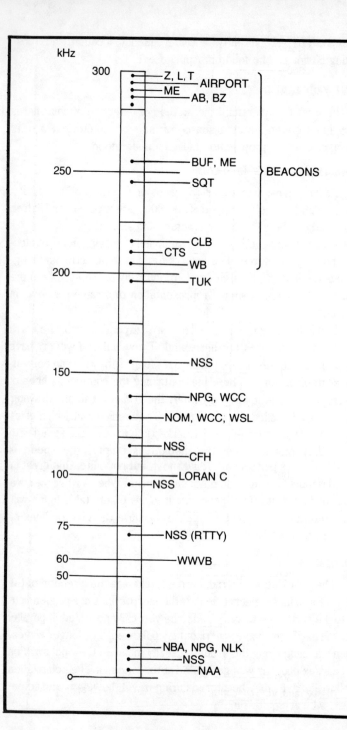

kHz

300 — Z, L, T
— AIRPORT
— ME
— AB, BZ

250 — BUF, ME
— SQT

— CLB
— CTS
— WB
200 — TUK

BEACONS

150 — NSS

— NPG, WCC
— NOM, WCC, WSL

— NSS
— CFH
100 — LORAN C
— NSS

75 — NSS (RTTY)

60 — WWVB
50 —

— NBA, NPG, NLK
— NSS
— NAA

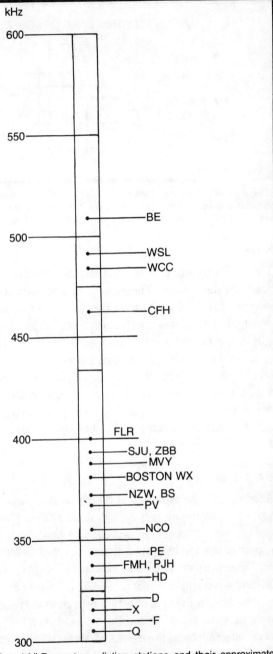

Fig. 1-1. The LF and VLF spectrum, listing stations and their approximate frequencies.

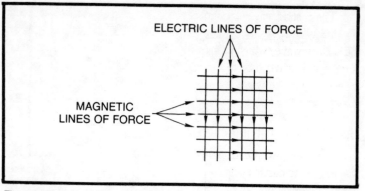

Fig. 1-2. Electric and magnetic lines of force in a radio wave. Arrows indicate instantaneous directions of the fields for a wave traveling toward the reader. Reversing the direction of one set of lines would reverse the direction of travel.

Spreading

The field intensity of a wave is inversely proportional to the distance from the source. Therefore if in a uniform medium one receiving point is twice as far from the transmitter as another, the field strength at the more distant point will be just half the field strength at the nearer point. This results from the fact that the energy in the wave front must be distributed over a greater area as the wave moves away from the source. This inverse-distance law is based on the assumption that there is nothing in the medium to absorb energy from the wave as it travels. This is not the case in practical communication along the ground and through the atmosphere.

Types of Propagation

According to the altitudes of the paths along which they are propagated, radio waves may be classified as ionospheric waves, tropospheric waves, or ground waves. The ionospheric or *skywave* is that part of the total radiation that is directed toward the ionosphere. Depending on conditions in that region, as well as upon transmitting wave length, the ionospheric wave may or may not be returned to earth by the total effects of refraction and reflection. The tropospheric wave is that part of the total radiation that undergoes refraction and reflection in regions of abrupt change of dielectric constant in the troposphere, such as may occur at the boundaries between air masses of differing temperature and moisture content.

The ground wave is that part of the total radiation that is directly affected by the presence of the earth and its surface features. One is the surface wave, which then is an earth-guided wave, and the other is the space wave (not to be confused with the ionospheric or sky wave). The space wave is itself the result of two components—the direct wave and the ground reflected wave as shown in Fig. 1-3.

PROPERTIES OF THE IONOSPHERE

Upon leaving the transmitting antenna, the wave travels upward from the earth's surface at such an angle that it would continue out into space were its path not bent sufficiently to bring it back to earth. The medium that causes such bending is the ionosphere, a region in the upper atmosphere, above a height of about 60 miles, where free ions and electrons exist in sufficient quantity to have an appreciable effect on wave travel. The ionization in the upper atmosphere is believed to be caused by ultraviolet radiation from the sun. The ionosphere is not a single region but is composed of a series of layers of varying densities of ionization occurring at different heights. Each layer consists of a central region of relatively dense ionization that tapers off in intensity both above and below.

Refraction

The greater the intensity of ionization in a layer the more the path of the wave is bent. The bending or refraction (often also called reflection) also depends on wave length; the longer the wave, the more the path is bent for a given degree of ionization. Thus low frequency waves are more readily bent than those of high frequency.

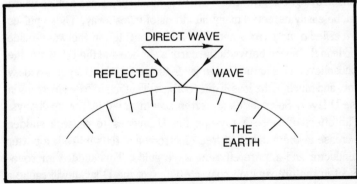

Fig. 1-3. How both direct and reflected waves may be received simultaneously.

Absorption

In traveling through the ionosphere the wave gives up some of its energy by setting the ionized particles in motion. When the moving ionized particles collide with others this energy is lost. The absorption from this cause is greater at lower frequencies. It also increases with the intensity of ionization, and with the density of the atmosphere in the ionized regions.

NORMAL STRUCTURE OF THE IONOSPHERE

The lowest useful ionized layer is called the *E Layer*. The average height of the region of maximum ionization is about 70 miles. The air at this height is sufficiently dense so that the ions and electrons set free by the sun's radiation do not travel far before they meet and recombine to form neutral particles, so the layer can maintain its normal intensity of ionization only in the presence of continuing radiation from the sun.

Therefore, the ionization is greatest around local noon and practically disappears after sundown. In the daytime there is a still lower ionized area, the *D Region*. D-region ionization is proportional to the height of the sun and is greatest at noon. It is this region that propagates VLF signals over long distances. The method of detecting radiation from solar flares can be done electronically. The effect of this radiation on the earth's atmosphere is very pronounced and this is particularly true of the lower part of the ionosphere at an altitude of about 70 kilometers.

A convenient VLF signal for flare detection is the natural noise caused by lightning. Each stroke of lightning produces a pulse which can be easily detected many hundreds of miles away. These pulses are called *atmospherics* and they are propagated in the wave-guide mode in the space between the earth's surface and the D layer of the ionosphere. This natural wave guide formed by the D layer is a lossy one, and much of the loss is due to a lack of sufficient free electrons in the D layer. Solar flares are often accompanied by strong X-rays. The effect of these X-rays on the D layer is to cause a sudden increase in the number of free electrons and thus make it a better conductor and a more efficient wave guide. This sudden improvement in the propagation characteristics of the D layer will cause a sudden enhancement of the level of atmospheric noise.

If a VLF radio receiver is receiving this atmospheric noise and recording its intensity on a strip chart recorder, the chart will show a sudden rise in level over a period of several minutes. This will be followed by a slow return to the normal level which usually takes from 30 to 90 minutes. One of these sudden enhancements of atmospherics, or SEAs as they are usually called, is shown in Fig. 1-4.

The second principal layer of the ionosphere is called the F Layer, which has a height of about 175 miles at night. At this altitude the air is so thin that recombinations of ions and electrons take place very slowly. The ionization decreases after sundown, reaching a minimum just before sunrise. In the daytime the F layer splits into two parts, the F_1 and F_2 layers with average heights of, respectively, 140 miles and 200 miles. These layers are most highly ionized at about local noon, and merge again at sunset into the F layer.

Cyclic Variations in the Ionosphere

Since ionization depends upon ultraviolet radiation, conditions in the ionosphere vary with changes in the sun's radiation. Very marked changes in ionization occur in step with the 11 year sunspot cycle. Although there is no apparent direct correlation between sunspot activity and critical frequencies on a given day, there is a definite correlation between average sunspot activity and critical frequencies. The critical frequencies are highest during sunspot maxima, and lowest during sunspot minima. The lower amateur frequencies 3.5 and 7 MHz frequently are the only usable bands at night. At such times the 28 MHz band is seldom useful for long distance work, while the 14 MHz band performs well in the daytime but is not ordinarily useful at night.

Ionospheric Storms

Certain types of sunspot activity cause considerable disturbances in the ionosphere and are accompanied by disturbances in the earth's magnetic field (magnetic storms). Ionosphere storms are characterized by a marked increase in absorption, so that radio conditions become poor. The critical frequencies also drop to relatively low values during a storm, so that only the lower frequencies are useful for communication. Ionospheric storms may last from a few hours to several days. Since the sun rotates on its axis once

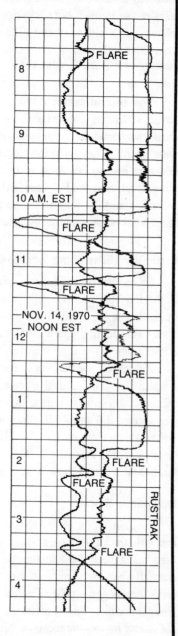

SEVEN SOLAR FLARES
DURING THE DAYTIME
NOVEMBER 14, 1970

LEFT-SIDE RECORDING IS
SEA (SUDDEN ENHANCEMENT OF
ATMOSPHERICS) AT 27 kHz

RIGHT-SIDE RECORDING IS
SES (SUDDEN ENHANCEMENT OF
SIGNAL) AT 24 kHz, STATION
NBA IN PANAMA

Fig. 1-4. An SEA recording.

every 28 days, disturbances tend to recur at such intervals, if the sunspots responsible do not become inactive in the meantime. Absorption is usually low and radio conditions good just before a storm.

METHODS OF DETECTING SOLAR FLARES AT VLF

There are two basic methods that can be used to monitor solar flare activity. The first method is by monitoring sudden enhancement of atmospherics using a long wave receiver to monitor the constant stream of static that seems to be most intense in the VLF range at about 27 kHz. The enhancement of atmospheric noise usually takes place every day between sundown and sunrise, whenever a solar flare occurs; however, the SEA can take place also during daylight hours. Whenever the radio signal shows a sudden increase or jump toward the nighttime readings, a solar flare occurs.

The second method is to use sudden enhancement of signal strength (SES). To use such a system a solid state receiver is designed to tune a VLF station of known origin, such as 18.6 kHz NLK at Jim Creek, Washington. The output of the receiver is recorded on chart paper. If the propagation path is on the daylight side of the earth during a solar flare, the ionosphere lowers and a sudden enhancement of signal occurs, causing a characteristic fast rise-slow decay curve on the chart recorder. Example of a full day's recording is shown in Fig. 1-5. Construction of both types of receivers will be outlined in Chapter 3.

THE AMERICAN ASSOCIATION OF VARIABLE STAR OBSERVERS

If you could watch the sun through a telescope continuously for several years you might eventually see a solar flare, one of the most violent eruptions of energy in our part of the universe. The flare would appear abruptly as a brilliant spot, persist for a few minutes and then fade out. Only the largest flares could be observed in this

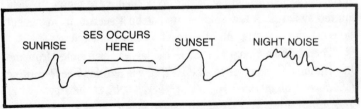

Fig. 1-5. What a full day's recording of SES might look like.

way. You can detect many smaller ones, however, if you have a radio receiver that can be tuned to very long waves. This part of the radio spectrum is usually quiet during the daylight hours, except of course on frequencies used by transmitting stations. During a solar flare the long wave receiver crackles with static of the type that we hear on the radio during a thunderstorm. In recent years this upsurge of radio noise has been used to detect the onset of flares. In the early part of 1960, David Warshaw of Brooklyn, New York, designed an inexpensive radio receiver expressly for the purpose of monitoring solar flare activity. When this receiver became available, a small group of amateurs across the nation volunteered to make a continuous patrol of solar flares as a contribution to the International Geophysical Year. The group's observations were published regularly in the Solar-Geophysical Data Bulletin of the National Bureau of Standards and distributed throughout the world to IGY World Centers.

The project actually got under way in 1955 when Harry Bondy, Chairman of the Solar Division of the American Association of Variable Star Observers, asked Mr. Warshaw to copy an elaborate radio receiver that had been designed in England for flare detection. Harry Bondy wrote that a group in the Solar Division was already providing sunspot observations to the Bureau of Standards for the computation of the American Relative Sunspot Number, an index of solar activity that is useful in making long range predictions of ionospheric conditions and is consequently of interest to the communications industry. The results of this work were well received in geophysical circles, so the group was on the lookout for a project that would enable them to participate in the IGY, which was then in the planning stage. One of the programs under consideration for the IGY was the study of solar flares. This program appeared to hold little promise for the group, because at that time the observation of all but the most intense flares required costly spectrohelioscopes or monochromators, instruments sensitive to certain wave lengths of light emitted by flares. A few enthusiasts had built such instruments but the equipment was beyond the reach of most amateurs.

Then in the summer of 1955 a circular was distributed from the U.S. National Committee for the IGY; it was entitled *The Recording of Sudden Enhancements of Atmospherics (SEA) for Purposes of Flare Patrol,* by M.A. Ellision, then astonomer of the Royal Observatory at Edinburgh. The circular explained how X-rays and ul-traviolet rays emitted by a flare cause a large increase in the iono-sphere at heights from 60 to 90 kilometers, and how this effect in turn greatly enhances the reflection of very long radio waves that obliquely strike the lowest of the ionosphere's four layers: the so-called D Layer. A proportional enhancement is simultaneously observed in the strength of radio signals received from a distant source after one or more reflections from the ionosphere.

According to Ellison, the effect was first described by the French physicist R. Bureau who investigated the enhancement of signals by recording the integrated level of tropical thunderstorm activity at many different wave lengths.

He found that sudden enhancements of atmospherics or SEAs are characteristic of the spectral range from 7,000 to 16,000 meters, being most pronounced at the wave length of 11,000 meters (27 kHz). At about this time, Mr. Warshaw, who was a member of the group and an electronic specialist in a communications company, built the first vacuum tube long wave receiver to monitor solar flares. Although the receiver performed well, the design proved to be impractical for amateur use. The original circuit called for costly British vacuum tubes and included a number of intricate biasing and voltage-regulating components to keep the indicating meter from fluctuating excessively during heavy signal bursts. After a short trial, Warshaw decided to develop a receiver of his own, utilizing transistors. The power and voltage requirements of the transistor are so low that batteries could be substituted for the conventional rectifier, eliminating the need for voltage regulation. The design that eventually emerged from Warshaw's basement workshop called for only three transistors, a diode, four coils, six resistors and capacitors, a pair of flashlight batteries and a microammeter. The circuit, including the batteries, was housed in a standard box, measuring only three inches wide, four inches deep and five inches long. Refer to Fig. 1-6.

Upon being tested, the new receiver exhibited excellent sen-sitivity. Flares of average size or above tripped a buzzer instead of being recorded on a pen recorder. On several occasions Warshaw found it possible to telephone his industrial colleagues that the cause of a current transmission blackout was ionospheric in origin and not an equipment failure. At this time Warshaw did not own a pen recorder and was relying on the buzzer to signal the onset of flares. The relay to actuate the buzzer was set to respond to all signals above a predetermined energy level. This arrangement quickly proved to be disappointing. The buzzer responded to all sorts of audio and noise, including interference originating in nearby oil burners, TV receivers, and electric shavers. SEAs differed from the spurious signals only in their rate of growth and decay that is in their *wave* shape. One way to identify them with reasonable certainty is to make a graph of the signal by plotting signal amplitude against time. This method required an automatic pen recorder. One day in August Warshaw located a secondhand recorder, and the next day he picked up a second one. By September he had the first instrument in working order. The first AAVSO Solar Division SEA Station was now in operation. C.L. Strong, another member of the group, began to analyze Warshaw's recordings. The hieroglyphics traced by the 27 kHz noise seemed utterly meaningless. The easily distinguished SEAs that stood out so clearly in the published data simply did not appear. Warshaw's station was located in Brooklyn at the intersection of Atlantic and Flatbush Avenues, where ignition noise is emitted by passing automobiles day and night. Disturbances from these sources appeared at first to mask all other effects. Finally on October 4th the first no-doubt-about it-SEA appeared on the recorder. In fact, the tracing for this data displayed three clear flares. These were immediately confirmed by Robert Lee of the High Altitude Observatory of the University of Colorado.

In the meantime, Walter A. Feibelman a physicist of Pittsburgh, Pa., became interested in the project and built a station after Warshaw's design. In May 1957 he submitted some of the clearest SEA's ever recorded. The second station was in operation. That Fall another member, Val Isham of Powell, Ohio, put a station in operation. The records submitted by these three stations compared favorably to those of professional observatories. Warshaw's design was vindicated.

After examining the results, Walter Orr Roberts, then Chairman of the U.S. IGY Solar Activity Panel, advised the group that the Bureau of Standards would assign four Brown recorders to the project for a year. Ellison, who had served as group consultant from the beginning, urged the patrol to set up at least two stations in high latitudes. There he thought SEAs of greater amplitude would be detected. The matter of putting your finger on competent amateurs willing to stick with the job around the clock for a year was no easy chore. There was also the matter of equipment maintenance. Some instruments were damaged in transit, others failed in use. In addition, there was always the problem of tracking down sources of local noise and taking corrective action. The recruiting job was handled by correspondence and after many exchanges four locations were selected for the recorders. It turned out that the chart drive of the Brown recorders run too fast for useful purposes, so the instruments were modified for a chart speed of one inch per hour. The recorders were then equipped with the 27 kHz receivers built by Warshaw and were shipped to stations in Oshkosh, Wisconsin; Manhattan, Kansas; and Edmonton, Alberta. Despite the best efforts of all concerned only one of the new stations contributed substantially to the flare patrol program. While these installations were on trial, C.H. Hossfield of Ramsey, New Jersey, assembled an apparatus of his own. Although situated well within the metropolitan area of New York City, this station made some of the clearest records submitted by the patrol.

What do SEA recordings show? Typically a 24 hour graph displays high amplitude signals during the night, followed by a low level trace beginning at sunrise. This characteristic is explained by the formation of the D Layer at sunrise. The long waves emitted by ever-present thunderstorms in the tropics reach the higher ionospheric layers by an unimpeded path at night and are reflected to the surface without significant loss of energy. During the day, however, the waves must traverse the lower and lightly ionized D Layer. Because of its light ionization, the D Layer acts not as a reflector but as a partial absorber. This loss of energy to the D Layer accounts for the precipitous dip in the recordings with the approach of sunrise. SEAs appear on the charts during the hours of sunlight because X-rays and ultraviolet rays emitted by solar flares impinge on the D Layer and increase the intensity of its ionization. The D Layer then acts as a reflector and the amplitude of recorded signals increases.

Proficiency in distinguishing SEAs from other disturbances comes with practice. The observer will acquire the knack by the time he has analyzed his first half mile of recording paper. Then the recordings can become exciting. When you spot a very large flare you can confidently predict a blackout in short wave radio communi-

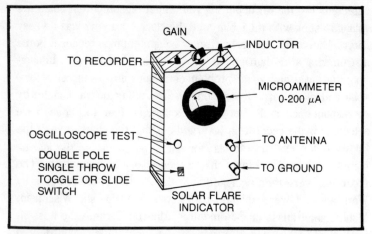

Fig. 1-6. Artist's illustration of the first solid state receiver used for the detection of solar flares.

cations within 26 hours, the interval required for the cloud of electrically charged particles ejected by the flare to reach the earth from the sun. By the Spring of 1957 all of the stations in the flare patrol were functioning routinely; however, one gross pattern evident in 80% of the records was the variation in signal intensity. The chart recordings dip at sunrise and rise with the approach of sunset.

A close examination of the recordings showed an interesting and equally regular pattern first observed in one sequence of recordings that spanned 14 days. These disclosed that the signal level does not drop abruptly with the approach of dawn. Instead the amplitude wavers for about 90 minutes, during which time four characteristic elements are observed. Some 45 minutes before the local sunrise the curve has a pronounced dip.

About 10 minutes later a typical hump appears. The patrol labeled this the precursor hump. The slump gradually slopes off with the approach of sunrise. The signal intensity then falls off to its daytime level. Finally, about 50 minutes after sunrise, occasional recordings show a post-sunrise hump of varying duration and intensity. What were the implications of the pattern? Although the question is still open, Warshaw has suggested one interesting explanation. The phenomenon could be accounted for by a detached zone of low ionization that forms at an altitude of about 50 miles on the dark side of the sunrise terminator (the line between the illuminated and dark side of the earth). This zone might be described as a detached portion of the D Layer. Signals reflected from a higher layer would be partially absorbed by the detached zone during their downward transit, an effect that would account for the pre-dawn dip. Then some 20 minutes before sunrise the earth's rotation would advance the detached zone three or four degrees and would again establish a clear path for the reflections. The precursor hump would now appear, as the signal intensity approached its nighttime value. The atmosphere in the detached zone may be ionized to about the normal intensity of the D layer by the ultraviolet rays that proceed through the atmosphere on the lighted side of the terminator and impinge on the detached zone at the optimum angle for creating ionization. The illustration in Fig. 1-7 shows the suggested mechanism together with an idealized graph of the sunrise pattern. No explanation has been advanced for the occasional post-sunrise hump. Its existence has been clearly established, however. On this side of the Atlantic the effect appeared most clearly in the records of the first flare station at China Lake, California. A word of caution, however, if you decide to undertake the task of the interpretation of flare recordings, you will discover that the post-sunrise hump can be vexing in the extreme. It resembles nothing so much as a clear SEA recording.

Usually the graph begins a gradual climb in the afternoon (the band becomes noisier). This is to be expected because the number of local thunderstorms, in addition to those in the tropics, increases during the latter half of the day. The signal pattern thus reveals its meteorological origin. Records show that relatively quiet a.m. periods exceed quiet p.m. periods by a ratio of 10 to 1. In spite of this, there is a strange excess of SEAs during the p.m. hours. Roughly 1.6 times as many SEAs—about 62 percent of the total— are recorded in the afternoon.

The flare patrol adopted the classification system for reporting SEA data that was proposed by J. Virginia Lincoln, Chief of Radio Warning Services Section of the Bureau of Standards, and which was described in Bureau of Standards Report 5540 (November 1957). According to this system the amplitude of SEAs ranges from 1 −, lowest in amplitude, to 3+, highest. Further they rated the certainty with which the identity of an SEA is established. This scale ranges from 5 (definite) to D (questionable). What good are the solar flare patrols recordings? First, they correlate satisfactorily with those made by official observations. They catch some of the SEAs

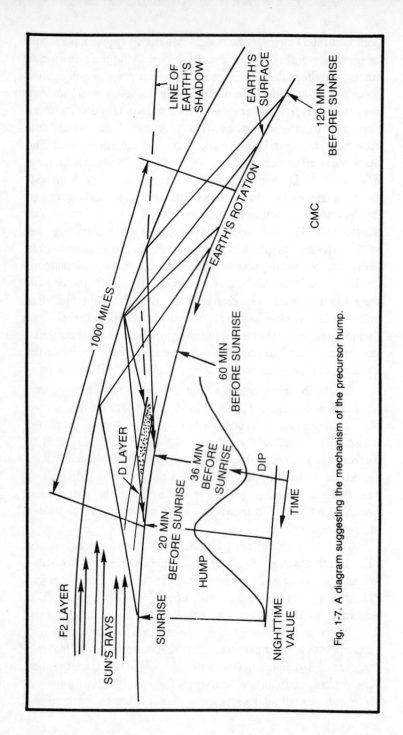

Fig. 1-7. A diagram suggesting the mechanism of the precursor hump.

28

that are missed by the official stations and vice versa. Today there are hundreds of flare patrol stations scattered at numerous locations throughout the world. The stations range from desolate outposts in Canada to ultramodern physics laboratories in colleges. There is always a need for additional observers and the Solar Flare Division of the AAVSO welcomes anyone who is interested in this type of solar flare activity.

IONOSPHERIC DISTURBANCES

The term *ionospheric disturbance* is used to cover a wide variety of ionospheric conditions that show some departure from the usual state. Thus slight perturbations in the electron configuration that move with time are called traveling disturbances. These are usually localized in space. From the point of view of radio transmission, these traveling disturbances are not nearly so important as the more (geographically) extensive disturbances which are all associated, in some way or another, with a flare on the sun. These solar flare associated effects may be classified as follows:

1. Sudden ionospheric disturbances (SIDs).
2. Ionospheric storms.
3. Polar cap absorption events (PCAs).

The reason these disturbances are important from the point of view of radio communications is that they often result in interruption of communications. The absorption in the D region is enhanced so much that radio communications may be impossible for periods lasting from a few minutes to several days. Furthermore, the critical frequencies of the F_2 layer are sometimes depressed (due to ionospheric storms), resulting in loss of signal. It is a very difficult matter to predict the onset of an ionospheric disturbance and since the ionospheric conditions can change rapidly, the idea of *average* conditions (e.g., path loss, maximum frequency) tends to have only limited value. Although all the regular layers (D, E and F) are affected, the E layer is not strongly affected except for the occurence of sporadic E in certain ionospheric storms. The D and F layers are subject to much stronger effects. The D region effects often occur simultaneously with the appearance of an optical solar flare, whereas the F region effects are often delayed by a day or more. These simultaneous and delayed effects are illustrated in Fig. 1-8. The occurrence of a solar flare may be accompanied by the emission

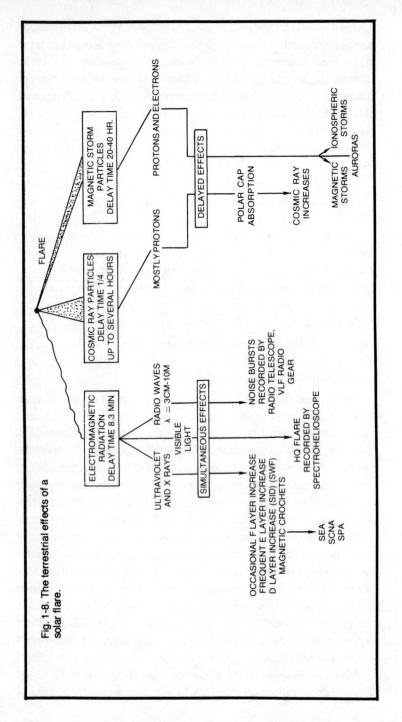

Fig. 1-8. The terrestrial effects of a solar flare.

of radio waves, ultraviolet, and x-rays, all of which arrive simultaneously at the earth because they travel with the free space velocity of light. Solar cosmic rays are also emitted and these may take 15 minutes to several hours to reach the earth, where they produce polar cap absorption. Slower particles, which have transit times of from 20 to 40 hours, result in ionospheric storms, magnetic storms, and visible displays of aurora borealis. It should be remembered, of course, that not all the effects shown in the figure are associated with every flare.

D-Region Absorption

At times communications of high frequencies by skywave propagation over the daylight hemisphere of the earth are "blacked out" by abnormally high absorption in the D region. The association between these shortwave fadeouts (SWF) and solar flares was discovered in 1936 by J. H. Dellinger. The condition of high absorption may last from a few minutes to several hours. Onset of this absorption is usually, but not always, very sudden (hence the name *sudden ionospheric disturbance,* SID); it is followed by a relatively slow recovery, as shown in Fig. 1-9(a). *Shortwave fadeouts* (SWFs) are sometimes accompanied by transient variations in the earth's magnetic field (Fig. 1-9(b)) indicating the existence of electric currents in the D region. Not all shortwave fadeouts are sudden, so at the National Bureau of Standards, they are described as sudden, slow, or gradual, depending on the time variation of the recorded signal. The production of ionization in the D region gives rise to associated phenomena such as the sudden absorption of cosmic noise (sudden change of atmospherics, SCA) received on frequencies above the F_2 layer critical frequency, the sudden phase anomaly (SPA) on very low frequencies, and the sudden enhancement of atmospherics (SEA) on very low frequencies. From the point of view of interruption of radio communications, shortwave fadeouts are not too serious, but they do constitute a nuisance. The following data on SWFs observed at Washington, D.C., gives an idea of this nuisance value. During the sunspot maximum year of 1937, SWFs were observed on 84 days. On 66 of these days the fadeout was classified as intense. On 39 days there was more than one SWF and on 33 days the fadeout lasted for more than 1 hour. On the other hand, in the sunspot minimum year of 1944, the corresponding numbers were of

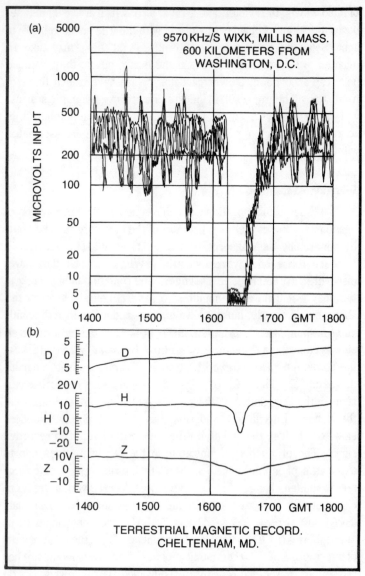

Fig. 1-9. Example of a shortwave fadeout and magnetic disturbance.

only 5, 3, 0, and 2. In the sunspot maximum year of 1947, they were again high—121, 104, 54, and 33. The height in the atmosphere at which the additional electrons are produced during a solar flare is of importance because it determines the dependence of the absorption on wave frequency. Fig. 1-10 shows the time variation of absorption of 10 MHz cosmic noise at College, Alaska, during a large solar flare. From the ratio of the absorption of the ordinary and extraordinary waves it is found that the frequency dependence is given by

$$(f \pm f_L) - 2.$$

On the other hand, there is evidence which indicates that absorption can also occur at low levels. It is of interest to note that during the flare shown in Fig. 1-10, the absorption in the D region increased by a factor of the order of 10. The vast majority of radio fadeouts is associated with solar flares. Therefore, they exhibit the same 11 years occurrence cycle as do flares and sunspots.

Effects on Very High Frequencies

Very high frequency waves are propagated via the D region by scattering from ionospheric irregularities. During shortwave fadeouts it has been found that these signals are enhanced by as much as 9 dB and there is little evidence of any weakening of signal (on 50 MHz) during an HF blackout. However, on 27.7 MHz, it was found that the signal enhancement lasted for a few minutes only and was followed by marked attenuation. From these observations it would

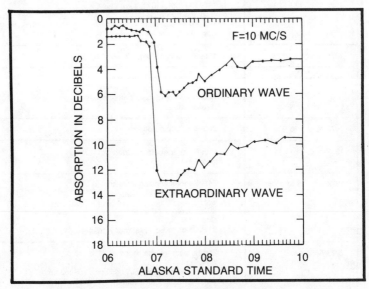

Fig. 1-10. Solar flare absorption on 10 MHz at College, Alaska.

Table 1-3. Typical Weather News Reports Coinciding With Recorded Solar Storms. Compiled by Ron Ham of Sussex, England.

Date of Solar Storm	Date of News Report
1970-Nov. 11 to 22	20th-East Pakistan flood disaster.
1970-Dec. 17 to 23	The unpredicted "White Christmas" in Europe with severe blizzards. (Some countries had snow for the first time.)
1971-Jan. 28 to 31	30th & 31st-Floods in Poland and Mozambique. Nine inches (22.9 cm) of rain in one day was reported from Australia. Heavy rain and snow in parts of U.K. The Thames River was at risk of flooding because of severe gales in the North Sea.
1971-Apr. 9 to 18	16th/17th-Worst weather in 72 years experienced by Mount Everest climbers. 13th-BBC news report that monsoons in E. Pakistan had started a month early.
1971-July 13 to 19	18th-Freak rainstorm in Seoul, Korea. 20th-Hong Kong hit by worst typhoon for many years.
1971-Aug. 18 to 27	26th-(Stop press). Storm havoc in Spain.
1972-Feb. 12 to 23	14th-100 mph (161 km/h) gales in southern France. 18th-Freak storms in Australia. 20th-Flooding in New York.
1972-March 3 to 12	13th-Flooding in Peru.
1972-June 15 to 22	22nd-Worst floods in American history; whole towns evacuated near New York; some parts declared disaster areas.
1972-Aug. 1 to 9	2nd-Flooding in London.
1972-Aug. 11 to 14	11th-Freak tornado reported in Holland.
1972-Oct. 19 to 31	24th-Serious flooding in Australia. 28th-Severe gales in Icelandic waters. 20th-Severe flooding in Costa del Sol. 31st-Gales in Somalia; many dead.
1973-April 1 to 11	2nd-Hurricane-force winds in Holland. 10th-Some parts of USA had snow for first time. 14th-Storms in E. Pakistan; many dead.

be concluded that the scattering occurred in the absorbing region and that the increased absorption was compensated for by increased scattering on the higher frequency, but not on the lower frequency.

Solar Activity and the Earth's Weather

Scientists for years have been trying to correlate the relationship between solar activity and the earth's weather fronts and storms. Ron Ham of Sussex, England, an amateur radio astronomer and a fellow of the Royal Astronomical Society, has operated a radio observatory from his home. He has correlated a relationship between solar storms and natural events. Through keeping records of solar and atmospheric events, Ham noticed that during or soon after he had recorded a noise storm the news media were likely to report (somewhere on earth) a freak, or violent, weather event, often with tragic loss of life, and/or extensive damage to property. The examples listed in Table 1-3 illustrate the typical weather news reports coinciding with recorded solar storms that set Ham thinking about a possible connection between these two natural events. These are just a few of the reports Ham has gathered from newspaper items and radio news broadcasts; there may well be many more, or more-detailed information in subsequent reports. Scientific literature tells us that a connection between the active sun and the earth's weather has been known for many years, but the precise link has not been yet identified. Briefly, certain changes in climate, and in plant life, have been associated with the 11 year sunspot cycle. For approximately 400 years, astronomers have systematically recorded the number of visible sunspots; and throughout these years, scientists have related many natural events to the existence of a large sunspot. The association of radio waves with an active sunspot has been recognized only for about 43 years. Prior to this, solar observations relied on what the eye could see. Could it be that Ham's observations identified the particular sunspot activity responsible for stirring up the existing weather systems on earth? After all, we know that solar particles which are heralded by radio noise, can upset the ionosphere; so why, by some indirect means, can't they upset the troposphere? Ham's theory was placed on record in an article in the Radio Society of Great Britain (RSGB) journal, *Radio Communication*, and today many scientists suggest that he provided the "lost" vital link in the chain of events between the sun and the earth's weather.

Chapter 2
Antennas

For the average amateur/experimenter, antennas that are usable for medium and low frequency *reception* fall into three basic categories. The first is a long wire, and depending on local conditions the longer and higher the better. The second is a vertical. This can be made from pipe or aluminum tubing, with an average height of 20 to 30 feet. The vertical should be properly guyed and set on a good base insulator. This type of antenna should work with an efficient ground rod at or near its base, which also can be used for lightning protection. A series of ground radial wires connected to the ground rod (like spikes on a wheel) will also improve the performance of the antenna. The third antenna is a resonant loop.

ANTENNA LOCATION

The location of your long wire or vertical antenna is *most* important. There is a severe receiving problem on the low frequencies, called radio frequency interference (RFI), and unfortunately most of it is manmade. Everything from appliances to welding arcs create manmade havoc to recorded signals, so the location of the antenna is most important. Fortunately most of these interferences are confined to a limited area. Keep all LF and VLF antennas away from your house wiring and power lines. A shielded lead-in, such as coax cable with the shield well grounded, will help reduce stray noise pickup. A small antenna that will not pick up noise is better than a large one that will. The secret to good reception is the readability of the signal over the noise level (signal-to-noise ratio). For some reason loop antennas have been ignored as a receiving aid by most hams, SWLs and experimenters, in general, yet almost every household contains one. There has not been an AM broadcast set made in the last 40 years that does not contain a loop antenna in one form or another. Most of these older sets have a common fault: they are directional, and, to the annoyance of many housewives, cannot always be placed on the table in a proper pattern to provide the best reception of a desired station. Usually they require a turn or two. The loop antenna is not a *gain device*; however, the null that it exhibits off its sides can be put to good use in attenuating (lessening) the strength of an interfering signal or noise source due to its direction pattern. In this chapter we will try to give you the best antenna design for the correct use.

RECOMMENDED ANTENNAS FOR SEA AND SES SYSTEMS

The following two antennas should perform well in receiving low frequency signals. The importance of a good ground system cannot be overemphasized.

The Maag Antenna

This antenna was designed by Russ Maag of Missouri Western State College in St. Joseph, Missouri. It is one of the earlier antennas recommended by the Solar Flare Network of the AAVSO. The antenna itself is 20 feet tall, made of two sections of 3 inch downspout tubing. This material usually comes in 10 foot lengths with slightly crimped ends for fitting together. Either aluminum or galvanized steel is suitable. The assembly is mounted vertically on a 2 × 4 wooden stake with its top shaped to prevent the downspout from slipping off. As the picture in Fig. 2-1 shows, the stake projects only a few inches above the ground, but a good 10 feet would be better. The wood can be treated with creosote to inhibit electrical conductivity in damp weather. Nylon guys spaced at 120 degree intervals keep the antenna upright. The feedline from the antenna to the receiver, or preamp if one is used, is regular six-strand copper antenna wire fastened to the middle of the downspout where the two sections couple together. A good, clean contact here is important. If a preamplifier is used (in weak signal areas), it can be

Fig. 2-1. The Maag antenna. Photo Courtesy Russ C. Maag.

mounted on a separate wooden pole 5 or 10 feet from the antenna so the signal is amplified before coming near any source of interference.

An excellent preamp in kit form is available from International Crystal and described in Chapter 6. For the very important ground, a piece of 1/2 to 1 inch stiff copper tubing driven at least 10 feet into the soil is ideal. Occasional watering of the ground near the tubing also helps conductance. The location of the antenna here is again important. Keep it away from existing power lines and away from power transformers.

The 27 MHz (CB) Quarter Wave Whip

The whip antenna is one of the simplest antennas of all. In fact it is nothing more than a single vertical rod (usually rather thin) made of a suitable conducting material and cut to a special length. A practical vertical whip may be almost any length up to 102 inches. A CB quarter wave antenna of this type can be fed directly with a 50 ohm coaxial transmission line, or one with a different impedance, provided it is used with a suitable impedance matching network. The CB fiberglass whips have been used with the solid state receivers, used by the Flare Patrol with excellent results. The whip can be mounted atop a suitable pipe or mast; however, the same grounding considerations should be observed as with other types of VLF antennas.

CONSTRUCTION OF A LOOP ANTENNA
FOR GENERAL LF AND VLF RECEPTION

This loop antenna, you will discover, is simple to construct, efficient and inexpensive. It has all of the desirable electrical features that a loop antenna should have. While the loop to be described was constructed for the author's RBA-7 low frequency receiver, which has a high impedance input, changes in the design and circuit have been made so that the antenna's feedpoint impedance is 52 ohms; therefore, it can be used with standard 52 ohm transmission line and easily matches the input impedance of most LF and VLF commercially available receivers. It may be used on other receivers having other than 52 ohm input impedance by the simple expedient of impedance matching. The decision by the author to construct a loop antenna for VLF in preference to using a random length long wire or a simple half wave dipole was made after realizing that such antennas have several undesirable characteristics. Assuming that the long wire was made long enough to have definite directional characteristics, it would be quite difficult to make use of the directional features because of the problem of positioning the antenna. The long wire is inferior to a loop antenna for minimizing noise pickup. A simple half wave dipole at 20 kHz would be approximately 4 miles long. Keeping the VLF antenna to a reasonable size and providing directional selectivity and noise reduction were factors that led to the choice of the loop. Loop antennas are not as commonplace as the "garden variety" of directional antennas, such as multielement directive arrays, for amateur radio and commercial use. Thus a few words regarding loop antenna operation may be in order for a better understanding of the construction details.

Loop Antenna Fundamentals

Loop antennas have been used for years in direction finding systems, particularly aboard aircraft and on vessels. The main function of the loop is to sense the direction of radio signals emanating from a transmitter at a fixed location. The very basic loop antenna is simply a coil of wire whose diameter is much smaller in comparison to the wavelength to which the coil is tuned. The ground wave type transmissions from LF and VLF stations cause vertically polarized waves to induce voltages in the loop wire as these waves pass by the loop. These induced voltages in the loop wire produce a loop current which depends upon the positioning of the loop antenna with respect to the wave front.

Almost any shape can be used for the design of a loop antenna, such as a square, triangle, octagon, diamond, or the conventional circle. Regardless of which shape the loop assumes, the maximum directivity is along the plane of the loop with a distinct minimum or null at right angles to the plane.

The directive pattern of a loop, whose diameter is small with respect to the wavelength to which the loop is tuned is similar to that of a doublet antenna, that is, a figure 8 field pattern. The minimum, or null, which is broadside to the plane, is extremely sharp and critical in a well designed loop antenna, and is normally capable of giving bearing information better than one degree in low frequency direction finding applications. While the purpose of constructing the loop antenna in this case is not that of direction finding, the presence of a sharp null at zero and 180 degrees with respect to a fixed LF or VLF signal source a reasonable distance away indicates that the loop is functioning properly. This null makes it possible to eliminate undesirable so-called adjacent-channel interference. The absence of this sharp null broadside to the plane of the antenna can be due to a number of improper actions such as locating the antenna too close to power lines, other antennas, gutters and downspouts, or other metallic objects. It can also be caused by poor symmetry of the antenna circuit, including the receiver input. The use of a balanced feedline and a pushpull RF stage for the receiver input—including a suitable matching transformer to match between a balanced feed point and an unbalanced line—can be accomplished to improve loop circuit symmetry. The presence of static electricity in the air is a source of much noise in low frequency reception and sometimes causes complete masking of desired signals. Enclosing the receiving loop wires in a nonmagnetic metallic shield will greatly reduce noise pickup thus enhancing the overall signal-to-noise ratio of the receiving system. The loop wires are completely surrounded by the shield except for a narrow transverse gap or break at the apex of the loop electrostatic shield.

Circuit Arrangement

The loop antenna that we are going to describe can be resonated from 14 kHz to 25 kHz using the components specified in the schematic diagram. If this range is used the feedpoint impedance will be 52 ohms. Construction of the antenna is quite simple and straightforward. The cost of the materials represents a very small investment for the performance obtained. The schematic of the VLF loop is shown in Fig. 2-2. Coil L1 is a continuous loop made up of 18 turns of 16 gauge enameled wire. L2, which is adjustable from 0.2 to 0.3 mH, is used to resonate the loop circuit to the desired frequency. Matching transformer T1 is wound so that loop balance is maintained while providing a match to a 52 ohm unbalanced line. Capacitors C1 and C2 are high quality micas used to bring overall loop tuning into the range of tuning coil L2.

The tuning coil, capacitors, and matching transformer are housed in a 5 × 7 × 3 inch aluminum box. The electrostatic shield for the loop wires is made from a 25 foot length of soft drawn, copper or aluminum tubing with a 1/2 inch inside diameter. Hard line coaxial cable with the center conductor removed can also be used. In many cases 25 feet of tubing represents half a standard (50 foot) length coil which can be used so that two loops result without any waste. A length of inexpensive plastic hose with a 3/8 inch inside diameter and slightly less than a 1/2 inch outside diameter was used

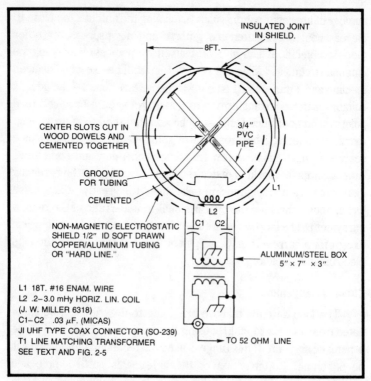

Fig. 2-2. Schematic of the loop antenna with its tuning and matching network.

inside the tubing to protect the loop wires during the pulling operation. The plastic hose also provides additional loop rigidity for the loss made necessary by the gap in the tubing at the apex of the loop.

Construction Details

The first step in construction of the loop proper is to uncoil the 25 feet of tubing (which was cut exactly in half from the 50 foot roll) on a level floor. The straighter you can get the tubing the easier it will be to pull through the plastic hose and wires in the following steps.

After the tubing has been straightened, and if you intend to use copper tubing, solder to each end of the 1/2 inch copper tubing a 1/2 inch outside thread adapter. If you intend to use aluminum tubing or "hard line," use an aluminum 1/2 inch tubing-to-thread straight connector. Attach the tube side to the end of the tubing by tightening up on the compression nut. Conduit nuts of suitable size (1/2 inch pipe) may be used to secure the copper tubing connectors to the

42

metal box housing the smaller circuit components. Next, measure to the exact center of the length of tubing and, using a tubing cutter, cut the tubing in half. Keeping the tubing sections together, insert a 27 foot length of plastic hose into one end of the tubing and, by working it slowly, pass the hose through both sections of the tubing so that approximately one foot of hose remains outside each end of the tubing. At this point a single 16 gauge wire should be worked through the tubing-hose combination to facilitate pulling through the bundle of 18 of these 16 gauge wires for the loop. The loop wires should be cut to lengths of 27 feet and each wire tinned on one end. The bundle of 18 wires is then soldered to the pulling wire and carefully pulled through the entire length of the loop tube. The tedious job of removing enamel insulation from the wire ends can be simplified by using a good grade of paint and varnish remover.

The easiest way to form the loop is to lay out a circle 8 feet in diameter on your basement or garage floor. Use a heavy marking pencil, so that the outline may be easily seen and easily removed. A word of caution, however. Do not construct the loop in a basement if you can't get it out of the doorway. For bending the tubing the author found that having someone standing on it prevented movement while the circle was being formed. Afterwards the plastic hose and wire should be cut back enough to allow mounting to the metal box. At this stage suitable holes may be drilled in the metal box and the threaded tubing ends fastened with conduit nuts. A terminal board is then installed in the center of the box for connecting the 36 wire ends into a loop of 18 turns. Identifying the wires can be accomplished with an ohmmeter or with a buzz-out circuit made from a dry cell and buzzer or pilot light.

If care is exercised the center turn of the multiturn loop can be identified and marked at the time that the loop wires are being soldered together to form the continuous coil thereby avoiding the trouble of trying to determine the loop center by electrical measurement. The loop wire resistance is low, in the range of 1.75 ohms, and with an ordinary ohmmeter, an accurate measurement is difficult. After the loop wires are all connected, with the exception of the two ends, tuning coil L2 can be mounted and connected in series with the loop wires at midpoint. Capacitors C1 and C2 can be mounted and wired except for the connection to be made to matching transformer T1. A suggested layout for the various components that are installed within the metal box is shown in Fig. 2-3. Line matching transformer T1 is an important part of the loop circuit in that it provides the impedance transformation required while maintaining loop balance. Reasonable care should be exercised in winding T1. The ferrite rod from a transistor radio loopstick was used as a core. The original windings were removed and 180 turns of 28 gauge enameled wire closewound on the core. This winding is the secondary and is used to connect the winding to coaxial output connector J1.

To determine how much wire will be required for the primary and how much space will be taken, wind 104 turns of the 28 gauge enameled wire over the secondary. Remove the 104 turns of wire and, after finding the center, fold it in half so that the wire is doubled back on itself. Then take the doubled wire and wind 52 turns over the 180 turn secondary. Reasonable care should be taken to center the primary winding over the secondary to maintain balance action. By connecting the wires of the primary as shown in Fig. 2-4 you will have a bifilar primary which will have equal capacitance from each end to ground. Coil dope can be applied to the windings of T1 to keep the wires in place. When this line-matching transformer has been

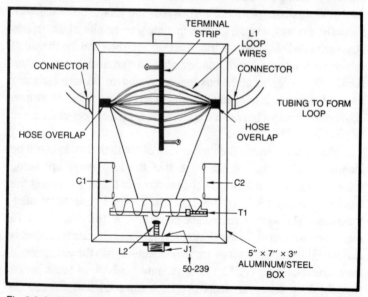

Fig. 2-3. Inside view of the metal box. The terminal board, fabricated from a piece of insulating material with holes drilled into it, can be used to support the connected ends of the 16 gauge loop antenna wires.

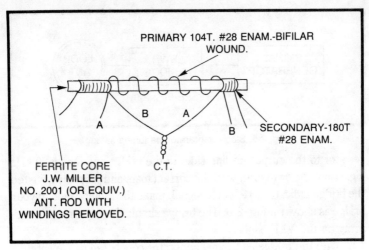

Fig. 2-4. Construction of the bifilar-wound balanced transformer.

completely fabricated it should be installed in the metal box and wired into the loop circuit. The loop antenna proper is now electrically complete and should be mounted in any convenient manner. The author used a combination of crosses made out of 3/4 inch outside diameter PVC pipe with wood dowels inside to construct the supporting cross members. Be careful not to use a solid metallic framework to support the loop since electrical operation of the loop will be seriously affected by shorting across the electrostatic shield. A piece of steel tubing attached to the lower part of the box supports the antenna. The tubing was then joined to a TV type antenna rotor to "steer" the antenna.

Tuning and Matching Impedance

Tuning the loop to resonate at any frequency in the 14 to 25 kHz range is relatively easy. It is accomplished in much the same way as tuning and impedance matching of higher frequency antennas. A simple block diagram of the setup is shown in Fig. 2-5. A standing wave ratio bridge that can be used is the Heathkit Model HM-15, or any other suitable HF unit. The audio signal generator can be any stable unit with a range of up to approximately 100 kHz such as the Heathkit IG-5218.

Before starting the adjustment procedure check for satisfactory operation of the bridge to be used on these very low frequencies. If the bridge is to check a 52 ohm termination, connect a 52 ohm

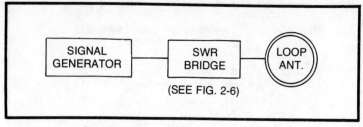

Fig. 2-5. Block diagram of the tuning setup.

resistor to the output, or line side, of the bridge and feed a 20 kHz signal into the bridge. If, with the correct load, and properly adjusted the bridge indicates a reflected signal, make the modification to your bridge as shown in Fig. 2-6. The bridge should then operate satisfactorily on the VLF band.

With the bridge and signal generator connected to the antenna as shown in Fig. 2-5 set the generator output to the frequency desired for loop antenna resonance. Adjust loop tuning coil L2 until the reflected power reading obtained on the bridge reads minimum. For the various input impedances which may be encountered it may be convenient to use one of the ultracompact high fidelity transformers manufactured by UTC or similar companies. The UTC type A-24 or A-26 has a response from 20 to 40,000 cycles with primary impedances of 15,000 ohms and 30,000 ohms, respectively. The secondary impedance for both of these transformers is 50, 125, 150, 200-250, 333 and 500-600 ohms, as required. By using either of these transformers in reverse it is possible to make a transformation

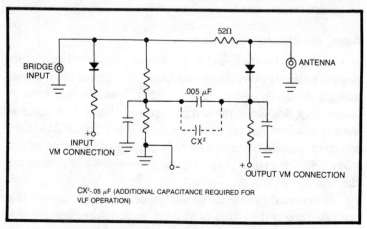

Fig. 2-6. An SWR meter modification to obtain VLF coverage.

between the low impedance 52 ohm line and the high input impedance of a receiver or converter. Another method in obtaining a higher feedpoint impedance for the loop antenna is to replace capacitors C1 and C2 and line matching transformer T1 with a single .015 μF capacitor to close the loop coil. When the loop is resonated by the use of tuning coil L2 you may tap at two points, one on either side of the electrical center of the loop and equidistant from the center. This will give a balanced feedpoint that is higher in impedance than 52 ohms. On the author's loop, tapping on the first two loop turns on either side of center gave a feedpoint impedance of 150 ohms. This impedance could be conveniently fed by the use of two 75 ohm coaxial cables with the shields tied together and the inner conductors connected, one to each loop tap. Since the loop turns are accessible in the metal box it is quite convenient to tap in this manner to find a satisfactory feedpoint.

Operational Checkout

Operation of the VLF loop antenna needs very little explanation. You will notice after a trial "run" that the loop exhibits two very sharp nulls broadside to the plane of the loop and that you have a much improved signal-to-noise ratio in the direction of the loop. Comparative checks can be made which will vividly demonstrate the superiority of the loop over the random long wire antenna.

A SHIELDED LOOP ANTENNA WITH PREAMP

The antenna that we are going to describe here was originally designed for WWVB reception, although it can be used for any low frequency application by simply tuning it to another frequency. Most of the LF and VLF commercially available receivers are fairly sensitive to low signals but with the use of this combination antenna and preamp the signals (especially WWVB at 60 kHz) will come through the speaker with S9 quality. To determine if this is the type of antenna needed at your location tune the LF receiver to 60 kHz. Using a vertical antenna kept away from all local metal, monitor the station one hour after dusk. Keep a close eye on the receiver's meter. If there is any hope at all, you should get a fairly strong signal on the meter with a distinctive once-each-second sudden drop in amplitude. You should be able to read the code, except for occasional noise pulses, and the background level should drop below the minimum signal as you tune off frequency. If you have problems receiving the signal this antenna is the solution to your problems.

Building the Antenna and Preamp

A shielded loop antenna is essential for the preamp. The antenna is shown in Fig. 2-7 and the complete preamp schematic in Fig. 2-8. Starting with 6 feet of 1/2 inch copper tubing, insert a piece of surplus 12 conductor *shielded* cable and bend it into a loop. A piece of *hard line* aluminum coax can be used although the center conductor and insulation must be removed. Terminate the cable in a suitable conduit housing that is big enough to hold the preamp. Be sure to use a plastic tubing fitting at one end to keep from getting a shorted turn on the shield. These are available in many hardware stores and are used specifically for shock proofing electric hot water heaters. The shield must be double (the cable plus the tubing) because the skin effect at 60 kHz requires considerable shield thickness. The final form of the loop will be slightly over 2 feet in diameter.

The loop is completed by wiring the conductors together to form a 12 turn loop, then soldering the shield and tubing at *one* end only. Resonate coil L2 to 60 kHz (or any other low frequency) with high-quality polystyrene capacitors or the much more expensive silver mica type. Any other type capacitor has proven to be unsuitable. The coil Q (for 60 kHz) should be around 25 to 40. More will cause temperature and possible tuning problems. Less will let in too much noise. Bearing these facts in mind, tuning the preamplifier to

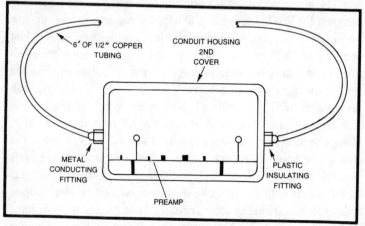

Fig. 2-7. A loop antenna with preamplifier.

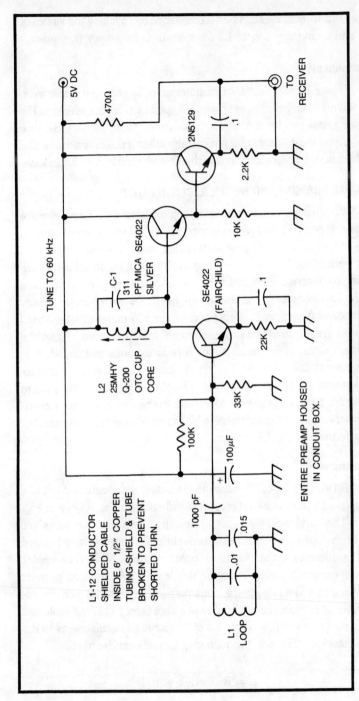

Fig. 2-8. Antenna preamplifier schematic diagram.

49

other low frequencies can be accomplished by changing the values of C1 and tuning the slug of L2 for maximum frequency response.

Components

The Fairchild SE4022 transistors can be obtained rather inexpensively from Poly Packs: consult Appendix B for the address. The silver micas can be substituted with good quality Mallory SX type polystyrene capacitors. Do not use any other type of capacitors. Coil L2 can be any good quality coil with a Q of 200 and 25 mH inductance.

A DUAL PURPOSE 160 METER RECEIVING LOOP

Construction of this simple loop antenna only takes about 4 hours. It serves a dual purpose, one as an ideal antenna for receiving the 160 meter amateur band and another as an experimental antenna for work in the low frequency region. By using the formulas we are going to describe here and the resonance tables given in standard radio handbooks, the antenna can be made to resonate on any number of frequencies. Reception of the 160 meter amateur band can be greatly improved by the use of a loop antenna. Again the primary virtue of the loop antenna is reducing noise pickup, which in this case is due to its relative small size and its directional characteristics when properly oriented. The loop antenna we are going to describe here is designed for 160 meters, although the experimenter can adapt the formulas to his likes and design an antenna that can be used in the LF region.

Construction

A two turn loop, 23 inches in diameter, was fashioned out of a length of 50 ohm jacketed coaxial semiflexible cable, as seen in Fig. 2-9. The vinyl jacket acts as a dielectric, and the two turns are securely taped, the ends being brought close together and bridged with sufficient capacitance to resonate at 1.835 MHz. The center conductor is ignored. A voltage divider across the ends permits coupling to a low impedance transmission line. The 23 inch dimension resulted from the availability of a short piece of scrap cable and was used as a starting point. To obtain a rough calculation as to the inductance of this loop the following formula can be used,

$$L = R^2N^2/(9R + 10S)$$

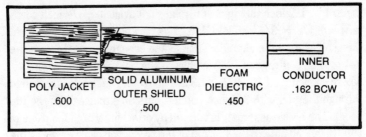

Fig. 2-9. Jacketed semiflexible coaxial cable.

Where L is the inductance of the coil in microhenries, R is the radius of the coil in inches, S is the length of the coil in inches, and N is the number of turns. For the two turn 23 inch loop, the calculation is as follows:

$$L = \frac{(11.5)^2 \ (2)^2}{(9 \times 11.5 + (10.1)} = \frac{529}{113.6} = 4.66 \ \mu H$$

Although this value may not be very accurate, it is quite helpful in determining an initial value of capacitance at C1 needed to resonate the loop to 1.835 MHz. Using the resonance table listed in The Radio Amateur's Handbook, this works out to around 1600 pF. By adjusting capacitor C1 it is possible to peak the response over a narrow portion of the 160 meter band. Tuning your receiver across the noise hump provides the most reliable indication of the resonance point. In Fig. 2-10 the resonating capacitance includes the parallel capacitance of the input voltage divider. Using the equivalent parallel capacitance as a basis (approximately 1424 pF), the actual inductance of the loop at 5.28 μH, is slightly higher than given by the formula at 1.835 MHz. A 1735 pF capacitor exhibits 50 ohms reactance. Values that are near to 1735 such as 1670 are satisfactory. A 100 pF capacitor can be used as a satisfactory coupling to the 50 ohm transmission line. Compared with receiving on a half-wave type dipole antenna, the loop performs remarkably well. Both the noise and strong signals are reduced by 20 dB or more, but more important is the fact that unreadable signals become clearly readable.

Mounting

The antenna can be mounted very simply and economically as shown in Fig. 2-11. After the loop is wound the two open ends are connected with plastic tubing connectors to a 3/4 inch PVC "tee." If the 3/4 × 1/2 inch tubing adapters are not available, 1/2 inch adapters with 1/2 × 3/4 inch PVC bushings may be used. Two wires are then soldered to the aluminum shield and brought down through the 3/4 × 6 inch threaded PVC nipple to the aluminum/steel minibox where capacitors C1, C2, and C3 are housed. The nipple is attached through the box with 3/4 inch conduit nuts on either side of the top panel. The entire assembly is rotated by attaching a 3/4 inch piece of PVC threaded pipe to the bottom of the box which also serves as the mast. For table top use the mast can be eliminated; however, a larger box should be used so that it can support the antenna and act as a base. The semiflex jacketed hardline is available at 45 cents per foot from Wire Concepts, Inc. Refer to Appendix B for the address.

SURPLUS LOOP ANTENNAS

Although loop antennas are not all that difficult to build, the serious low frequency experimenter should always be alert for good buys on the surplus market. These antennas usually offer heavy duty

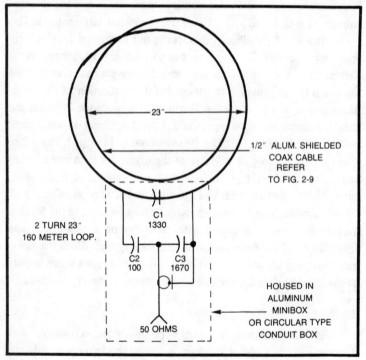

Fig. 2-10. Coupling and resonating a two turn loop.

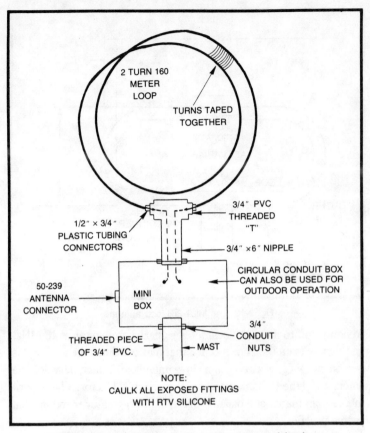

Fig. 2-11. Assembly details for the 160 meter receiving loop.

mechanical construction and electrical characteristics usable as is or with slight modification.

The MLA-2/B Loop Antenna

The MLA-2/B Loop Antenna is finding its way to surplus dealers and electronic outlets throughout the country. This antenna was designed by Antenna Research Associates of Beltsville, Maryland. Figure 2-12 is a drawing of the MLA-2/B, a rather odd looking loop device that has the capability of handling a kilowatt (when used in the transmit mode) continuously over the high frequency band from 3 to 25 MHz. The antenna can be put to a number of uses by the amateur/experimenter.

First, it can be used as a standard low frequency loop antenna, with minor modifications. Second, it can be used by the radio as-

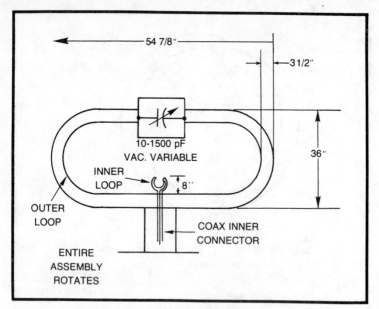

Fig. 2-12. The MLA-2/B loop antenna.

tronomy buff to monitor decameter signals from Jupiter at 18 MHz, and signals from the sun, too, for that matter. Since these systems rely on large signal strength, a close impedance match between the antenna and the transmission line is not a critical factor. The experimenter can match and balance the system to whatever extent that seems practical.

The MLA-2/B does not exhibit a great deal of gain except above 17 MHz, but the antenna has an excellent elevation pattern that covers both high angle and low angle radiation, with a 4 MHz coverage that is almost perfectly uniform as seen in Figure 2-13. The horizontal pattern is the typical figure eight of a loop, and it offers excellent directivity, with an inherent noise figure improvement over a dipole. The pattern roll is rated as sharp as 20 dB. The MLA-2/B's longest suit is its size, offering the performance of a half-wave antenna with a device no longer than 54 inches. The design of the antenna could be a starting point for some very useful amateur/experimenter projects. From a commercial standpoint the manufacturer has a patent on the antenna, but amateurs and experimenters alike might use it as a basis for their own experiments. The theory behind the MLA-2/B is an impedance transformer from a 50 ohm transmission line to the nominal 377 ohm radiation resistance of free space. The input to the antenna is to a feed loop placed within the primary loop. The signal is inductively coupled to the primary, which is the radiating body. There is no direct connection between the primary (outer) loop and the feed loop. Tuning of the primary loop is accomplished by varying a motor driven, remotely controlled vacuum variable capacitor inserted in the top of the loop. This value is varied from 10 to 1,500 picofarads. The loop itself is constructed of 3-1/2 inch diameter tubing, and the whole assembly is a rugged device, capable of standing up in 100 mile an hour winds. The feed loop is matched to the transmission line by bringing the coax into one side of the loop, as per the ratio of the diameter of the feed loop to the primary is 1:6. In the actual MLA-2/B the feed loop appears to be more or less a standard aircraft loop direction finder about 8 inches in diameter, very similar to the old MN-26 used in World War II. The primary loop has an average radius of 22 inches, but it is flattened, and the resulting shape is 51-7/8 inches across (center of tubing) and 36 inches high (center to center of tubing). The MLA-2/B needs to be operated between 3 and 5 feet above the surface and requires no ground system. The author has not used the loop for transmitting purposes; however, it has been tested on a receiver and its performance was excellent over a wide range of frequencies, directions, and wave angles.

The AT-382 Loop Antenna

The AT-382 loop antenna is an ideal "table top" low frequency antenna measuring only 9 inches in diameter. It was intended for use

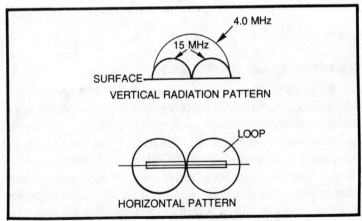

Fig. 2-13. Coverage pattern of the MLA-2/B loop antenna.

Table 2-1. Specifications of Commercially Available Tracor VLF/LF Antennas.

	599-805	MODEL 599-818	599-825
Frequency Coverage	10 kHz to 30 kHz (VLF)	10 kHz to 14 kHz (OMEGA)	10 kHz to 100 kHz (VLF/LF)
Mast Height	20 feet, unguyed constructed of three telescoping sections	10 feet, unguyed	20 feet, unguyed constructed of three telescoping sections
Equivalent Height at 20 kHz	0.024 meters	0.010 meters	0.024 meters
Construction Material	Stainless Steel	Stainless Steel	Stainless Steel
Termination Resistance	50 ohms	50 ohms	50 ohms
Recommended Cable	RG-58/U or RG-58A/U	RG-58/U or RG-58A/U	RG-58/U or RG-58A/U

with LF receivers in the 190–1500 kHz range. For its small size the AT-382 exhibits rather surprisingly good sensitivity and noise free reception. Overall dimensions are 14 × 9 × 4 inches. Weight averages around 10 pounds. The loop comes complete with a stand, and requires a small spline flex shaft and control box for remote tuning. To put the unit in operation, the only procedure required is to replace the UG-88 antenna connector with a standard SO-239 type.

COMMERCIAL LF AND VLF ANTENNAS

Although the market for commercially manufactured LF and VLF antennas has not been very profitable for companies (except for military contracts), a few do manufacture antennas that are used in conjunction with their VLF instrumentation. These antennas can be used with any commercial LF or VLF receiver and offer the amateur or experimenter who is not a "constructor" an opportunity to purchase a ready made antenna at moderate cost. One such company is Tracor Instruments. Tracor manufactures a complete line of vertical antennas that are either all band, or band pass units. The antennas are well constructed and require little or no maintenance since the elements are made of stainless steel. All of the antennas have 50 ohm termination impedance, and use standard connectors. A listing with specifications of the Tracor antennas can be found in Table 2-1.

Tracor also manufactures a complete line of long wire antenna couplers that are designed to match a long wire antenna to a 50 ohm cable feeding any LF/VLF receiver. Further information on either of these products can be obtained by writing to the manufacturer listed in Appendix B.

Chapter 3
Building Solar Flare Receivers

A very common method used for the detection of ionospheric disturbances is to tune a long wave receiver to a low or very low frequency
signal source and record its intensity, or amplitude, on a strip chart
recorder. Two basic receiving systems are used.

DETECTION METHODS

The first, or oldest, as we have discussed in Chapter 1 is called
the Sudden Enhancement of Atmospherics (SEA) method.   When
using the SEA receiving method, a radio receiver is tuned to either
24 or 27 kHz. The signal source is the natural radio pulse (static)
caused by lightning. With thousands of thunderstorms in progress at
any one time around the earth, there is a relatively constant signal
source of pulses being received through the ionosphere from great
distances at both 24 and 27 kHz. By recording the amplitude of these
pulses on a chart recorder, disturbances due to solar flares will result
in an increase in the amplitude of the received signal, usually rising
rather suddenly, then decaying slowly as the ion clouds dissipate
back to a normal state which usually takes from 30 to 90 minutes.

The second receiving method is quite similar to the SEA
method. It is called Sudden Enhancement of Signal (SES). A known
radio station either LF or VLF can be used as a signal source.
Receivers with frequencies between 17 and 100 kHz are usually
used to detect the SES type disturbances, with theories indicating

that the ionosphere tends to be most responsive at frequencies nearer to 24 and 27 kHz. The following comments are given to supplement the circuit diagrams and photographs shown in this chapter. They apply to the construction of three basic solid state receivers. The first is a combination SEA/SES type receiver, utilizing a linear IC and a single transistor, that can be tuned to 24 and 34.5 kHz. The second construction project describes a deluxe version of the same receiver that is tuned to 18.6 kHz (station NLK located at Jim Creek, Washington). This receiver uses a ceramic filter for selective operation, and is built on a printed circuit board. This receiver should be the choice if the amateur/experimenter lives in a very noisy area.

The third project is the construction of a four transistor SES receiver and preamplifier that can be designed to operate on any number of frequencies, such as 17.8, 18.6, 21.4, 22.3, 24, and 36.2 kHz.

A BASIC SEA AND SES RECEIVER FOR 24 AND 34.5 kHz

This dual frequency receiver, as the schematic diagram in Fig. 3-1 shows, is completely solid state and uses a linear IC for the amplifier. It has been used by a major Pennsylvania college for over 10 years with excellent results. This receiver can be housed in almost any type of wood or metal enclosure. The entire circuit, with the exception of the power supply, can be mounted on a piece of Vectorboard. Coils L1 and L2 are mounted 1/2 inch apart and their tops are epoxied into 1/4 inch holes in the Vectorboard. The linear IC (RCA 3035) should be mounted in a socket for ease of assembly and also for replacement purposes although it has a very long operational life. Pin 8 of the linear IC should be grounded (independent of the other grounds in the circuit) using a heavy conductor such as 12 or 14 gauge wire. This is critical since grounding is an important part in the successful operation of a flare receiver. Capacitor C2, switch S1, and bias potentiometer R4 are external controls and should be mounted on the front panel of the enclosure. If you intend to use the "open frame" type enclosure, they can be mounted directly on the Vectorboard. Wiring of the circuit (with the exception of the grounding bus on Pin 8 of the IC) is not critical; however, good component location and soldering procedures should be used. Transistor Q1 is soldered directly into the circuit. Extreme care must be

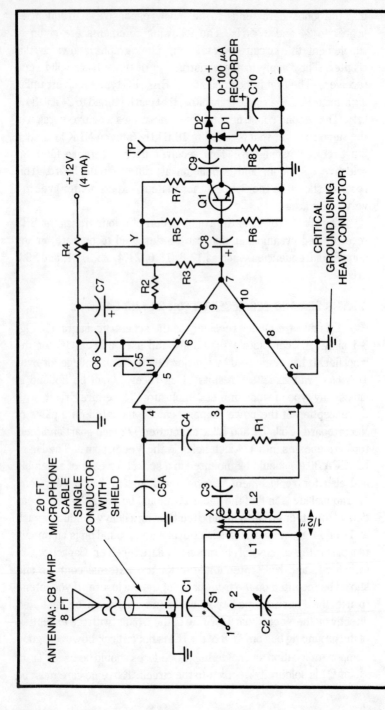

60

C1, 3, 4, 5, 9 0.001
C2 ARCO (ELMENCO) 308
C5A 0.005 μF
C6 0.05 μF
C7 50 mF ELECTROLYTIC
C8, 10 10 mF ELECTROLYTIC
R1 220K
R2 1K
R3 16K
R4 5K POT
R5 80K
R6, 8 10K
R7 2.5K
S1-S.P.D.T.

FOR 34.5 kHz, S1 AT POSITION 1
FOR 24 kHz, S1 AT POSITION 2
L1, 2 MILLER # 6319
L1 OPEN 13 TURNS
L2 OPEN 24 TURNS
AT Y, 7.5V MAX
Q1-SK3019
      SK3018
U1 RCA CA 3035

* DESIGNATES PARTS MOUNTED
   ON FRONT OF CABINET, IF
   USING ENCLOSURE.

E   CASE   B
          C
                SK 3019
                SK 3018

Fig. 3-1. Schematic diagram of the SID receiver for SEA/SES at 24 and 34.5 kHz.

exercised when soldering the transistor leads. Each lead should be held with a pair of needle pliers to act as a heat sink to conduct heat away from the transistor. The 20 foot length of microphone/antenna input cable can be substituted for with a similar length of 52 ohm coaxial cable, either RG-8 or RG-58. The cable length should be no longer than 40 feet; however, if greater lengths must be used, a "gain" type vertical antenna is recommended. Another option is to use standard 8 foot CB whip with a preamplifier mounted at the antenna.

Fig. 3-2 shows the receiver constructed on a printed circuit board instead of a Vectorboard. Capacitor 2 was eliminated and replaced for fixed frequency control; however, the parts orientation is the same as that used with Vectorboard. This type of "open frame" construction is accomplished by sandwiching the upright board in a groove between two pieces of 6 × 5-1/2 × 3/4 inch pine stock, using a 5-1/2 × 4 inch piece for the back support. Screws are used to hold the top and bottom pieces together. This open frame type of construction is easily built, and inexpensive.

Fig. 3-2. Photograph of the SID receiver used for SEA/SES studies, showing the open frame method of construction.

Part Substitutions:

One of the problems that the amatuer/experimenter frequently encounters in building projects is the availability of parts. Although this receiver uses standard electronic components, obtaining some of these parts may be difficult in some areas; therefore, we have listed some substitutions.

- RCA-CA3035 Substitute Sylvania ECG785
- RCA-SK-3018 Substitute GE-39
- RCA-SK-3019 Substitute GE-11
- Arco 308 If necessary, order this capacitor directly from Arco at the address given in the Appendix.

Tuning and Operation

Tuning the receiver to operate at 24 and 34.5 kHz is a relatively simple manner. Miller coil L1 is opened 13 turns, coil L2 is opened 24 turns. Gain control R4 is set fully counterclockwise and a 12 volt regulated dc power supply is attached to the power input. A CB 8 foot whip antenna is then attached to the antenna input jack. The only test gear required is a VTVM and a 0– 200 μA or 0– 50 μA meter, such as those used for monitoring signal strength on receivers. An alternative method is to use a VTVM and oscilloscope to monitor signal level. Either method is acceptable.

Tuning Using VTVM and 0-200 μA Analog Meter. A 50 μA meter can also be used in this procedure but strong signals will peg it before a peak is reached.

1. Attach the 0-200 μA meter to the recorder output terminals (observe correct polarity).
2. Set VTVM for measuring 12 volts dc and attach the probe to point Y.
3. Set switch S1 to position 1.
4. Gradually increase R4 to 7.5 volt maximum.
5. Re-peak coil L1 for maximum signal 10 to 50 μA or more depending on the strength of the incoming signal.
6. Set switch S1 to position 2.
7. Repeat steps 4 and 5 (re-peak coil L2).

Note: During this procedure capacitor C2 may have to be opened 1/4 turn.

Tuning Using VTVM and Oscilloscope. A sensitive meter of about 200 μA or a strip chart recorder will be required to verify operation.

1. Attach the oscilloscope probe to the TP connection on board.
2. Set the VTVM for measuring 12 V dc and attach the probe to point Y.
3. Set switch S1 to position 1.
4. Gradually increase R4 to obtain 7.5 volts, maximum.
5. Re-peak coil L1 for maximum signal on oscilloscope.
6. Set switch S1 to position 2.
7. Repeat steps 4 and 5 re-peaking coil L2. After tuning is completed, connect either a 0-100 μA strip chart recorder or a 0-200 μA meter to the recorder output terminals. Observe either the meter or the recorder for a noticeable up and down shift of signal. If this occurs the receiver is operating properly.

A DELUXE SES RECEIVER FOR 18.6 kHz

This receiver, as we had mentioned before, is an ideal project for the amateur/experimenter plagued with noise problems such as highpower lines and other types of manmade interference. This receiver, as the schematic diagram in Fig. 3-3 shows, is almost identical to the previous receiver except for three important changes. The first change is the elimination of variable capacitor C2, substituting fixed capacitance in its place. The second change is the addition of an 18.6 kHz ceramic filter to the circuit for selective operation. The third change can be seen by viewing Fig. 3-4 which shows that the entire receiver is built on a circuit board.

One of the outstanding characteristics of the Vernitron LF ceramic filter used in this receiver is its phase stability. It is constant at the center frequency within 1° or less through a 60 dB dynamic range. What this means is that the filter is able to "lock" a signal in on center frequency. The unit comes packaged in a HC/6U can and can be soldered directly into the circuit board. The same construction procedures should apply to this receiver as those used for the previous receiver. Since all of the controls and terminal strips are mounted on the front part of the circuit board this receiver is ideally suited for an "open frame" type of enclosure.

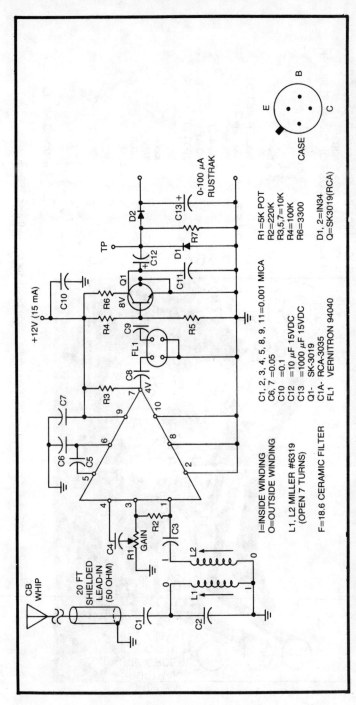

Fig. 3-3. Schematic diagram of the SID receiver for SES, modified to receive 18.6 kHz, station NLK at Jim Creek, Washington.

65

PRINTED CIRCUIT BOARD
SES (SUDDEN ENHANCEMENT OF SIGNAL)
RECEIVER-IC RCA CA3035
MODIFIED FOR USE OF CERAMIC FILTER

L1,2   MILLER #6319
C1,2,3,4,5,8,9,11   .001 MICA
C6,7   .05 MICA
C10   .1 MICA
C12   10 μF
C13   1000 μF
R1   5K POT
R2   220K
R3,5,7   10K
R4   100K
R6   330

D1,2   IN34
Q   SK3019 (RCA)

F CASE   CERAMIC FILTER

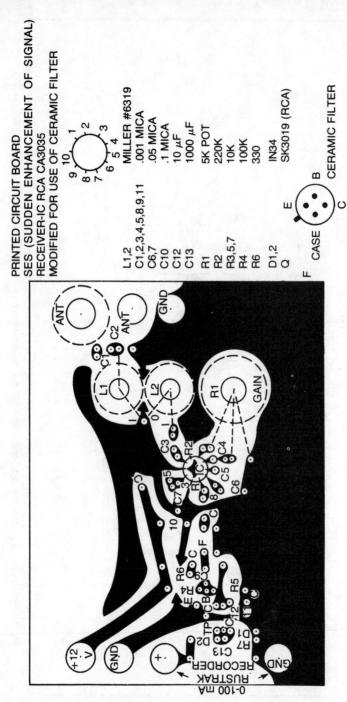

Fig. 3-4. Etching and drilling guide for laying out a circuit board for the 18.6 kHz SES receiver.

66

Tuning and Operation

This receiver can be tuned either with an analog meter (0-200 μA) or an oscilloscope. The same tuning procedures apply as to receiver 1.

Adapting to Other Frequencies

This receiver can be modified to tune any number of low frequencies. This may be the preferred method of operation for two reasons. First, you may want to get the receiver "on the air" as soon as possible after construction is completed, and the Vernitron ceramic filters have rather long lead times. Second, you may want to construct a receiver with a minimum of money invested. This receiver can be adapted to other SID frequencies without the ceramic filter by following these tuning procedures:

1. Place a jumper wire between capacitors C8 and C9.
2. Run the core of L1 and of L2 completely in.
3. Turn the gain control 1/4 turn clockwise.
4. Connect an oscilloscope probe to the test point and a ground.
5. Connect the antenna and ground the receiver.
6. Turn L1 out one turn. A large sine wave should appear on the oscilloscope. The magnitude should be rather large. This is 17.8 kHz coming from Cutter, Maine.
7. Continue to turn the core of L1 slowly out. The signal will drop and a small "hump" will appear on the screen. This will be 18.6 kHz.
8. Continue to turn the core of L1 until a rather large signal again appears. This is 21.4 kHz.
9. Continue to turn the core of L1 until a very small signal appears on the screen. This is 23.4 kHz from Hawaii.
10. Continue to turn the core of L1 out until at least 1/2 of the core is out. A large signal will appear again but not nearly of the same amplitude as seen at 17.8 kHz and 21.4 kHz. This will be 34.5 kHz.
11. Disconnect the oscilloscope and connect the signal into the recorder. Turn the gain up to around 70 or 80 μA. Fine tune the signal by observing the recorder and turning L1. You will see at what point the signal goes up and down. Peak the signal at 34.5 kHz.

12. Fine tune the signal with L2. In some cases adjusting it may cause an appreciable increase, in others it may not (10-12 turns). Since each coil is not the same, tuning of L2 may vary.

13. To test for unwanted oscillation, disconnect the antenna. The signal should drop to zero, and there should not be any noise.

If these tuning procedures are followed carefully, the 34.5 kHz signal should be the preferred choice for monitoring solar flares, although the experimenter may want to monitor some of the other frequencies for comparison purposes. The tuning should be done at night or after sunset when the noise factor is at a minimum. Complete drilled and etched circuit boards for this receiver can be obtained from the author. Consult the Appendix.

A VARIABLE FREQUENCY VLF RECEIVER

This receiver is unique in that it can easily be made to operate on any frequency in the VLF range. The first receiver of this design was built in July of 1976 to receive signals from a VLF station operating at 34.5 kHz. As the schematic diagram in Fig. 3-5 shows, the receiver can be tuned to other VLF station frequencies, such as 24.0, 23.4, 22.3, 21.4, 18.6, and 17.8 kHz, simply by adding more capacitance to C1 and C3 to let them resonate at these frequencies. Resistor R1 is a high value used to let atmospheric electricity drain off to ground. Resistance is noncritical, as anything from 1 to 20 megohms will do. Capacitor C2 minimizes the effect of antenna capacitance on the input circuit. That is, it minimizes loading on tuned circuit C1 and L2. C1 can be anything from 1000 to 2000 pF without much change normally. (This depends on your individual antenna and its effect on the tuned circuit.) Changing the capacitance of C2 will require tuning of the slug in L1 to obtain resonance at the desired frequency, but if you want to experiment and make C1 quite large, some method of checking resonance of C1 and L2 must be available to be sure it will tune as high as the desired frequency. Capacitor C1 is 240 pF for 34.5 kHz but it would probably be necessary to use 500 to 700 pF to tune from 22 to 24 kHz. The simplest way to tune this resonant circuit is to connect a 100 kilohm resistor in series with the output of an audio generator and attach this signal source to the antenna input of the receiver. Then with a

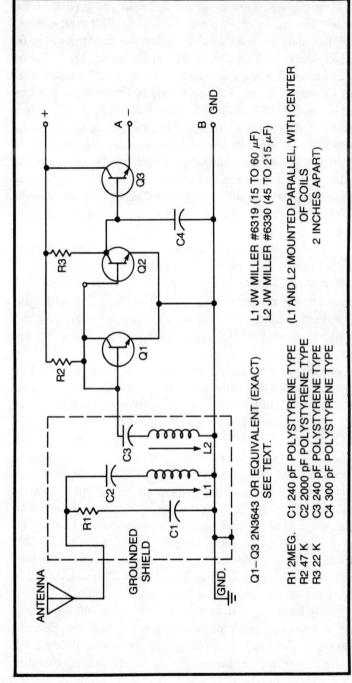

Fig. 3-5. Schematic diagram of the three transistor SES preamplifier circuit that can be designed for use on a number of frequencies.

L1 JW MILLER #6319 (15 TO 60 μF)
L2 JW MILLER #6330 (45 TO 215 μF)

(L1 AND L2 MOUNTED PARALLEL, WITH CENTER OF COILS 2 INCHES APART)

Q1–Q3 2N3643 OR EQUIVALENT (EXACT) SEE TEXT.

R1 2MEG.   C1 240 pF POLYSTYRENE TYPE
R2 47 K   C2 2000 pF POLYSTYRENE TYPE
R3 22 K   C3 240 pF POLYSTYRENE TYPE
             C4 300 pF POLYSTYRENE TYPE

VTVM or oscilloscope also across the input and ground of the receiver, find the resonant point of the C1 and L2 parallel tuned circuit by tuning the audio oscillator across the band from 20 to 60 kHz. Resonance will result in a high indication on the VTVM or scope. Turning the slug in L1 one turn will tune to about 34.5 kHz. Coils L1 and L2 are mounted by drilling two holes about 2 inches apart in a 3 × 4 × 5 inch Minibox, used as a shield. Push the screw of each coil into its hole in the box. The hole spacing is not critical; however, if they are mounted too close, excessive coupling occurs and tuning may be very broad although good transfer of energy will take place. If the coils are spaced more than 3 inches apart, there will not be enough coupling and the signal level will be low. The receiver itself can be constructed on either Vectorboard or on a printed circuit board.

Inductor L2 and capacitor C3 form another resonant circuit that must be tuned to the desired frequency. This is a series tuned circuit. It can probably best be tuned at the same time that L1 and C1 are tuned by turning up the gain control (R4 in Fig. 3-6) about 1/4 the way and either using a $0 - 200$ μA μετερ ορ α'' − 200 μA chart recorder as an indicator. Before turnign on the power to the receiver (12 volts), check that the output of the audio generator is not set too high as it will cause the meter or recorder to go ff scale very quickly. Set the output of the audio oscillator so the recorder or meter indicates that a small signal is being received. Then adjust L2 to resonate at the desired frequency. It will probably be necessary to reduce output of the audio generator as resonance occurs to keep the recorder on scale. After this initial adjustment of tuning slugs on L1 and L2, it may be necessary to fine tune both coils to the desired signal by disconnecting audio generator and just using the radiated signal and antenna on the input. Peak both coils to the received signal, keeping the meter or recorder on scale with the gain control. The value of C3 for 34.5 kHz is 240 pF. This will also tune to 24 kHz but that is about as low as frequency as can be reached without adding more capacitance. Capacitors C1 and C3 should be either polystyrene or mica type capacitors.

Following the input stage are three direct coupled transistors, really one small amplifier of the IC type but built with discrete components. This type of amplifier circuit was used in this receiver because experience has proven that it is very unlikely to suffer damage from nearby lightning strokes, a problem that plagues most LF circuits. Also, this circuit is not damaged easily by incorrect connections. It does have the drawback, however, of not operating if other types of transistors are substituted—even though some cross reference manuals indicate substitutions are possible. Although the author suspects several other types could be used, the only ones that have proven to work well together are 2N3643 and 2N2712.

Capacitor C4 is a 270 to 300 pF mica or polystyrene capacitor to roll off gain at higher frequencies and to help eliminate local broadcast station interference. Resistor R4 is the gain control network, and can by anything from 2,000 to 3,000 ohms. Capacitor C5 is a voltage blocking capacitor and any value from 0.01 to 0.4 μF will perform well.

INTEGRATOR CIRCUIT

This second amplifier for the receiver, shown in Fig. 3-6, is a one transistor circuit able to use nearly any small-signal type transistor. If you are familiar with transistor circuits, you may note that there is no resistor and bypass capacitor in the emitter lead to the transformer. The reason for this is that we used no emitter/degeneration of the transistor in the integrator so that a substantial increase in gain can be gotten out of the circuit; therefore, you can trade off higher amplification against increased distortion. Distortion at this point in circuit does not distract from the operation as it would in a high fidelity circuit. Audio transformer T1 is one of the inexpensive, common transistor types. While we used a 2K to 10K impedance anything similar would be suitable to use such as 1K to 10K, 1K to 20K, or 2K to 20K. Be sure to connect the smaller impedance to the collector of the transistor, which is backwards from standard use. This is done to raise the voltage, not to match the impedance. Doing so helps to overcome the 0.3 volt forward conduction breakdown of the diodes so that the "knee" of the curve has less effect on the output to the recorder. The diodes used in this circuit are germanium. *Do not use silicon*—not that they would not work, but they have a higher forward breakdown voltage, so the output to the recorder would be less linear at the lower end of the scale. Capacitor C6 is an integrating capacitor that gives an integration time of about 4 or 5 seconds when used with a Rustrak or Simpson 0 to 100 μA recorder with a 4600 ohms meter movement. (Chapter 4 discusses

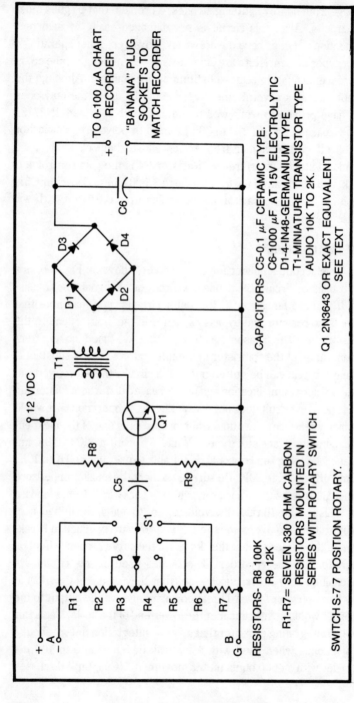

RESISTORS- R8 100K
R9 12K

R1-R7= SEVEN 330 OHM CARBON
RESISTORS MOUNTED IN
SERIES WITH ROTARY SWITCH

SWITCH S-7 7 POSITION ROTARY.

CAPACITORS- C5-0.1 μF CERAMIC TYPE.
C6-1000 μF AT 15V ELECTROLYTIC
D1-4-IN48-GERMANIUM TYPE
T1-MINIATURE TRANSISTOR TYPE
AUDIO 10K TO 2K.

Q1 2N3643 OR EXACT EQUIVALENT
SEE TEXT

TO 0-100 μA CHART RECORDER

"BANANA" PLUG SOCKETS TO MATCH RECORDER

Fig. 3-6. Schematic diagram of the integrator circuit used in conjunction with the three transistor preamplifier.

72

these recorders in detail.) Most professional observers use an integration time of 10 seconds, but the author prefers a shorter period. Integration time is the 1/e time, or time that it takes a capacitor when combined with a resistance and/or inductance, to charge or discharge about 63 percent. This capacitor may be viewed as a device that smooths the signal output and reduces the spikes or valleys in the signal.

Preamplifier and Integrator Parts Data

Resistors, capacitors, diodes, and the transformer are available from any radio-electronics outlet such as Radio Shack or Lafayette Electronics. The J.C. Miller coils are available through Lafayette Electronics and various other radio supply houses; however, the number of places carrying Miller products is decreasing each year. The coils can be ordered direct from J.C. Miller. Their address appears in the Appendix. The 2N3643 and 2N2712 transistors should be available from local supply houses.

Experiments in transistor substitution to an HEP or RCA number have not been too successful. The doping must be proper to work in the preamplifier section since the first transistor sets bias for the system. This three transistor amplifier has a low impedance input and output and a common one-transistor circuit cannot be used, without reducing the input impedance to about 100 ohms, because the L2 and C3 sharpness of tuning is dependent on the input being low to the amplifier. *Beware of transistor substitutions*. If the receiver is going to be designed for frequencies lower than 24 kHz, coils L1 and L2 can be Miller 6330s, with higher inductance.

Preamplifier Mounting

The preamplifier section of this receiver, shown schematically in Fig. 3-5 is housed in a small can mounted near the mast and electrically connected to the upright (vertical) water downspout antenna by open lead wire, as Fig. 3-7 shows. The can is grounded and electrically isolated from the vertical antenna by the use of nylon bolts or other nonconducting material. The bottom of the can should contain a terminal strip so that the coaxial feed cable can be attached using spade lugs soldered to the end of the cable wires. The outer weatherproof jacket of the cable should be bound to the box or antenna downspout with metal or plastic strips and self-tapping

Fig. 3-7. Preamplifier section of the SES transistorized receiver showing the method used for attaching the amplifier, housed in an aluminum can, to the antenna mast and side of a building. Photo Courtesy Russ Maag.

screws to anchor the cable, thus preventing breakage due to "wind slip." The conductor cable to be used here can be any good grade of two conductor and braided shield. The shield braid should be soldered to a spade lug for attachment to the ground post of the barrier terminal strip. The shielding can or "bud" box should also be grounded at the same point by a heavy piece of copper or aluminum wire. This type of coaxial cable may be obtained from several manufacturers. A good economical, weatherproof type is obtainable from most Radio Shack stores and sells for $1.59 for a 20 foot coil. Belden cable, and the RG-50 series cables that are used for TV coaxial service are excellent since they use the two conductors with a ground shield.

Integrator Mounting

The integrator circuit shown schematically in Fig. 3-6 can be contained in a small Bud enclosure and placed in the observatory,

74

house, or any other building where the actual recording is to be done. The preamplifier circuit, previously discussed, is coupled to this circuit by use of coax cable. The actual hookup of the coaxial cable leading from the preamplifier section to the integrator section should be made by using Jones type connectors. Jones connectors come as a matched male-female pair. The points +, −, and G are connected to matching terminals of the male-female sections of the Jones plug. This makes an interconnection arrangement that is easily disconnected, and, since these connectors are keyed, also easily reconnected.

Optional Gain Control Circuit

The schematic diagram in Fig. 3-8 shows an optional gain control circuit consisting of a 2.2 kilohm potentiometer, which may be substituted according to one's preference. The gain control section shown in Fig. 3-6 consists of seven 330 ohm resistors mounted on a rotary switching. This step response provides a better gain control for a builder of this receiver interested in "getting into the network" of solar flare observers. Especially in the United States this step response will provide a particular trace on recorder paper that can be better matched with others having similar recorders. The step gain control has definite values whereas with the potentiometer the value is continuously variable, thus completely unknown.

Power Supply

While this receiver is designed to work on 12 Vdc, the circuit should work well from 10 to 15 volts but with 15 volts giving better

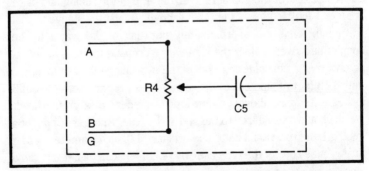

Fig. 3-8. Optional gain control circuit for use in the flare network AAVSO system.

75

gain. The use of a regulated power supply is recommended; however, in most localities the change in voltage should not change the gain enough to affect the performance of the receiver. If voltage does change, shifting the recording trace enough to be evident, the use of a regulated power supply is recommended.

There are numerous power supplies on the market that can be used with any of the forementioned receivers. One such is a kit available from ADVA Electronics. It is continuously variable from 2 to 15 Vdc, has a typical regulation of 0.1 percent, electronic current limiting at 300 mA, and is easily assembled in one hour. Refer to the Appendix for the manufacturer's address. An alternative would be to use two 6 volt batteries in series to obtain 12 volts.

Grounding

Perhaps *the* most important single factor in building this LF transistor receiver is that the total system be adequately grounded. A good grounding rod is a section of 1/2 to 1 inch diameter copper tubing. This copper tube should be driven into moist ground to a depth of 18 feet! Copper tubing is readily available in various lengths and wall sizes at plumbing and hardware stores. An easy method of acquiring a deep ground is to obtain 18 feet of copper tubing, the "hard" type used for household water lines. Cut this into 1 foot sections. Also purchase 17 sleeve couplers of the proper size for the tubing diameter. Drive the first 1 foot section into the ground by placing a board over the top of the tube and hammering the tube into the ground by gentle blows. The copper tubing is very malleable and easily damaged, so care must be exercised in this process. Drive the section of tubing into the ground leaving about 3 to 4 inches sticking out. Place one of the couplers over the end of the tubing and place another 1 foot section of the tubing above. Adjust the sleeve so that the tubing ends are approximately one half of the way into the coupler and sweat solder the joint with a propane torch. By successively driving and soldering each of the remaining tube sections, the entire 18 feet of pipe can be driven into the ground. Leave about 3 inches extending above ground for attachment of a ground lead, which should be soldered to the end of the pipe. Another option for a good ground rod would be to use ordinary 1/2 inch diameter, 4 foot long ground rods of the type used by telephone companies and radio amateurs. These can be welded together to make a 16 or 20 foot grounding rod. These rods, obtainable at electronic or radio and television supply houses, are copper coated iron.

The soil area around the grounding rod *must* be kept moist at all times. "Watering" the area with a garden hose is perhaps the simplest method. Even in the winter time when the temperatures are near or below freezing the area must be kept moist.

The Antenna

The *Maag* antenna described in Chapter 2 is perhaps the best and easiest to construct for this type of receiver.

Antenna Coupling

Coupling to the Maag antenna takes place at the center of the downspout, where the two 10 foot sections are telescoped and bolted together. Use RG-59/U coaxial transmission cable. This is the same cable (72 ohms) used to couple a TV antenna to the input of the TV receiver. This type of cable has *one* isolated center "hot" copper conductor and an outside conductor made up of a braided sheath. At the antenna hookup point *only* the "hot" center wire is attached under one of the bolt heads or self tapping screws. This provides good contact with the "downspout."

The cable should then be mechanically secured by a small metal band wrapped around the sheath and downspout, just for protection from "wind whipping." In other words, firmly support the cable at the downspout. After attaching the "hot" center lead, this screwed down attachment point should be heated by a propane torch and solder allowed to thoroughly flow into and around the joint. The braided shield of the cable should be left hanging unattached, but, of course, trimmed smooth with a pair of electrician's side cutters or diagonal pliers. The other end of the center conductor of this cable is then attached to the junction of R1 and C2, as shown in Fig. 3-5. This is located in the shielding can. Then the center conductor of the cable leading from the can at point C3 of the schematic is run to the integrator circuit by whatever length of cable is necessary to couple the antenna to the inside of the building where the integrator is kept. Again the braided shield is left "floating" unattached to anything. The 18 foot ground pipe at the base of the antenna is the *only* place the system is grounded.

Tuning and Operation

It is really not necessary to use an audio oscillator, oscilloscope, and frequency counter to tune to 34.5 kHz if you are located in North America since a very powerful LF station on this frequency is situated in the central United States. The following procedure decribes the most simple procedure for tuning the receiver-integrator.

1. Apply power to the receiver after the antenna and ground system are attached.
2. Connect a 0–100 μA recorder or meter as an indicator for determining when the signal is best tuned in. If using a recorder, make sure the stylus is free to move.
3. Turn the slug of coil L2 (Miller 6330) so that 16 turns or 16 threads are showing on the screw, including the threads at the end with the screwdriver slot. This should give the approximate frequency if the same components are used as shown in the diagrams.
4. Set the gain control to near maximum, observing the recorder or meter readings to keep an on scale indication.
5. Adjust L1 to give maximum signal indicated on the recorder or meter. For the circuit shown in Fig. 3-5, this was 14 turns out; however, this will vary considerably with individual antenna characteristics.
6. Fine tune L2 and L1 several times to peak the signal. While doing this tuning, it will probably be necessary to reduce the gain several times to keep the recorder or meter on scale.

If you are east of Utah in North America, it is likely that you have tuned the receiver to 34.5 kHz. This can be verified by recording the signal for 24 hours, noting the effects of sunrise and sunset. There is also an LF station on 37.2 kHz in Mohave, California, that could be tuned in by mistake, although this station operates at considerable lower power outputs, and it is not easily received in the eastern United States. If you are located west of Utah, it is recommended that you employ the use of an audio oscillator or oscilloscope to verify the correct frequency. These radio stations are occasionally off the air for maintenance purposes, so tuning should be delayed a few days and tried again if a station is not found the first time. The sunset effect, occurring 20 to 40 minutes after local sunset, and the

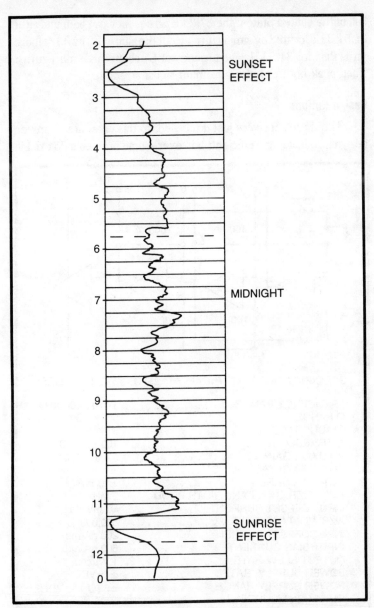

Fig. 3-9. Typical trace from a solar flare recorder showing sunrise and sunset effects.

sunrise effect, occurring 20 to 40 minutes before local sunrise, as shown in Fig. 3-9, are periods of low signal strength which should be avoided when trying to tune any LF or VLF receiver. In the western half of the United States, the 37.2 kHz frequency can be found after 34.5 kHz is located by simply turning out the slug of L2 by 3 1/4 more turns than for 34.5 kHz and turning out L1 about three more turns. Then peak both coils for maximum signal strength.

Design Options

The three receivers just discussed, to this date, are of proven designs, reliable, and efficient; however, some problems do exist in

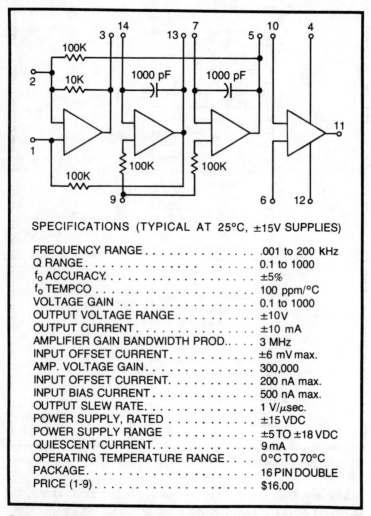

SPECIFICATIONS (TYPICAL AT 25°C, ±15V SUPPLIES)

FREQUENCY RANGE	.001 to 200 kHz
Q RANGE	0.1 to 1000
f_o ACCURACY	±5%
f_o TEMPCO	100 ppm/°C
VOLTAGE GAIN	0.1 to 1000
OUTPUT VOLTAGE RANGE	±10V
OUTPUT CURRENT	±10 mA
AMPLIFIER GAIN BANDWIDTH PROD.	3 MHz
INPUT OFFSET CURRENT	±6 mV max.
AMP. VOLTAGE GAIN	300,000
INPUT OFFSET CURRENT	200 nA max.
INPUT BIAS CURRENT	500 nA max.
OUTPUT SLEW RATE	1 V/μsec.
POWER SUPPLY, RATED	±15 VDC
POWER SUPPLY RANGE	±5 TO ±18 VDC
QUIESCENT CURRENT	9 mA
OPERATING TEMPERATURE RANGE	0°C TO 70°C
PACKAGE	16 PIN DOUBLE
PRICE (1-9)	$16.00

Fig. 3-10. Schematic diagram of the Datel model FLT-U2 universal active filter including specification sheet.

areas of high noise and interference. The main source of this problem relates back to the coils used in the antenna tank input circuit. Although shielding is recommended, along with a ceramic filter as used in the deluxe SES receiver, some stray interference may plague the recordings, giving pseudo traces appearing as solar flares. One such way to eliminate this problem would be to eliminate the coils and substitute for them a frequency active filter. Such a filter can also be used as a primary selectivity element. Large fixed value inductors are normally used to achieve selectivity at audio frequencies. A more modern approach would be to use a resistor-capacitor network combined with an amplifier to synthesize the characteristics of an inductor. When this "inductance" is resonated with a capacitor, an audio tuned circuit called an *active filter* results. Datel Systems manufactures a universal active filter, model FLT-U2 that falls in the frequency range of LF and VLF specifications (from .001 to 200 kHz). It uses the stated variable active filter principal implemented with three committed op amps, resistors, and capacitors, and, as Fig. 3-10 shows, a fourth uncommitted op amp which can be used to provide summing, buffering, gain, or an additional pole. The filter provides two pole lowpass, highpass, and bandpass functions simultaneously and can be used in phase correction and notch circuits. The filter is tuned by four external resistors which set the gain, center frequency, and Q of the circuit. The Q range is up to 1000 and the resonant frequency range is up to 200 kHz. Resonant frequency accuracy is typically ±5% with good stability. The units come packaged in a 16 pin double spaced ceramic DIP. The serious amateur/experimenter will find this filter a healthy challenge to experiment with, perhaps to initiate the design of a noise free, stable LF or VLF receiver. It can be obtained from electronic distributors or directly from Datel. Consult the Appendix for the manufacturer's address.

AN IC LF RECEIVER

Although solar flares have not been recorded above 76.8 kHz, the experimenter may want to build a small solid state receiver to monitor some of the Canadian localized weather forecasts, or low frequency broadcast stations operating between 150 and 300 kHz. The answer to this is the British made Ferranti Electric ZN414 radio receiver integrated circuit. This device contains practically all com-

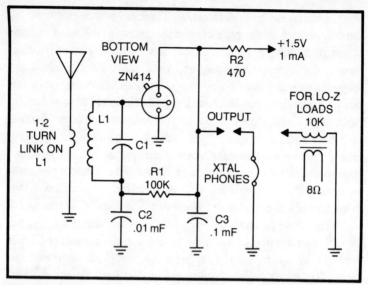

Fig. 3-11. Hookup of the Ferranti ZN414 radio receiver chip. R1 is the agc resistor.

ponents and circuitry for a 10 transistor TRF receiver in a small TO-18 3 lead package. The ZN414 provides a complete AM radio circuit which operates from 1.1 to 1.8 volts and requires only a battery, headphones, and antenna plus a tuning capacitor, coil, two

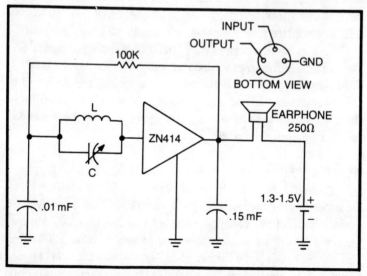

Fig. 3-12. Complete receiver, usable from 150 kHz to 3 MHz.

decoupling capacitors, and two resistors. Effective built-in agc is provided which is variable if required. Sufficient output (typically 30 mV rms) to drive a simple audio amplifier is provided at a total current consumption of only 1 milliampere. Excellent audio is achieved with a total harmonic distortion of less than 1 percent. Figure 3-11 shows the schematic of the circuit. The experimenter will have to choose LC values for the frequency range of interest. The ZN414 will function from 150 kHz to 3 MHz. To obtain good selectivity the inductor and capacitor chosen should present a Q of 70 to 100 or so to obtain selectivity comparable to that of a superhet.

Specific precautions for circuit layout are as follows:

1. Capacitor C3 should be as close to the chip as possible.
2. All leads from the ZN414 should be as short as possible.
3. A ferrite antenna coil, if used for L1, should be at least one inch away from the ZN414 to prevent oscillation.
4. The value of R2 can be selected for best agc action.

In Fig. 3-12 we see another version of the receiver. Here again LC values should be selected for the frequency range of interest, and the Q of the combination should be as high as possible. Signals down to 30 μV or so at the input of the chip will be receivable

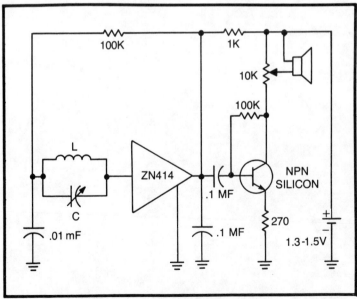

Fig. 3-13. Amplified version of Fig. 3-12.

83

with this circuit. In Fig. 3-13 one stage of amplification with a volume control has been added for greater convenience. Also a crystal earphone is used instead of the magnetic one. Other circuits and interesting applications as well as performance data for this unique integrated circuit are given in a brochure published by Ferranti. Also available from the same source are the ZN414s.

Refer to the Appendix for further information.

Chapter 4

Recording The Signals

The easiest and most economical way to graphically plot the output of the receivers that we discussed in Chapter 3 is by the use of miniature strip chart recorders. The most widely used are the Rustrak Model 288 and the Simpson Model 2750. These galvanometric stylus driven recorders are easy to use, require minimal service, and can be obtained either new or used-reconditioned. A new recorder sells in the one to two hundred dollar range while a used recorder can be purchased for about one-third of that. An excellent feature of this type of recorder is that they can run continuously for a period of 31 days without changing paper. The recorders are so designed that, by the use of interchangeable gear trains, a number of chart speeds can be installed to fit your choice. These recorders use the inkless method of recording using sensitized paper, giving a neat, smudge free continuous trace.

THE RUSTRAK (GULTON INDUSTRIES) MODEL 288 RECORDER

The Rustrak Model 288 is a miniature automatic strip chart recorder, as shown in Fig. 4-1. It features a sensitive galvanometer and a moving strip of pressure sensitive chart paper. Every two seconds a horizontal striker bar clamps back the meter pointer to record its position on the chart scale. This dry marking process is smudge proof and eliminates messy inks, ribbons, or carbon. The recorder can operate in any position. Normal chart speed is 1 inch

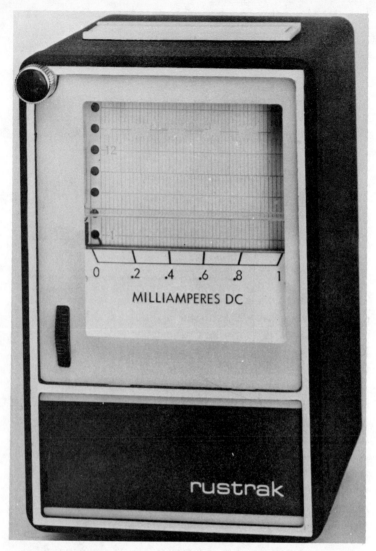

Fig. 4-1. Rustrak Model 288 miniature strip chart recorder. Photo courtesy Rustrak Instruments, Gulton Industries.

per hour for 31 days of continuous recording with a standard 63 foot roll of chart paper. The chart strip is driven through a time drum by a 115 V ac, 60 Hz, synchronous motor and gear train. Other chart speeds can be obtained simply by changing gear trains. The response time of the recorder is rated at 1 second maximum. Rated accuracy is two percent of full scale, more than adequate for solar flare recording. Critically dampened, overshoot is 2 percent of full scale. The safe working voltage is rated at 750 volts, maximum peak value. Meter scale width is 2 5/16 inches. The overall size of the recorder in inches is: 3 5/8 W × 5 5/8 H × 5 1/8 D. Other features of the recorder include optional reroll and tear off capability, an access window for marking, a thumb operated chart advance, and a quick chart review. When ordering either a new or used Model 288 recorder the following specifications should be included with the order: Model 288 Recorder, range 0-100 μA, and an internal resistance of 4700 ohms. Either a 110 Vac or 12 Vdc drive motor can be specified depending on the power requirements available. The recorder can be ordered (new) through any of the Gulton Industries representatives throughout the United States. Refer to the Appendix for a listing of new and used recorder outlets.

Wiring

All plug connections and terminals on a new Rustrak recorder are identified on a label inside the recorder; however, on used recorders, in most cases, this label has been removed or is not readable. The schematic in Fig. 4-2 shows the wiring and terminal connection diagram of the Model 288 recorder, both for a 110 Vac and a 12 Vdc motor drive hookup. The option of using a 12 Vdc drive motor instead of 110 Vac is left up to the experimenter. To build a total noise free recorder station away from all ac power sources, the use of the 12 Vdc drive motor operated from gel/cell batteries is strongly recommended. These dc drive motors can be ordered new from Rustrak in a variety of motor speeds. The 2 rpm version is the standard speed used for the 1 inch per hour chart drive and is the recommended speed used for the receivers described in Chapter 3.

Paper Loading

If the recorder is ordered new, it comes complete with paper loading instructions; however, from past experience, a used recorder probably won't have instructions or wiring diagrams with it. Figure 4-3 showing an exploded view of the Rustrak 288 recorder, should be studies before any attempt is made to load the recorder.

Reroll Method. The italic numbers in the following steps refer to item numbers in Fig. 4-3.

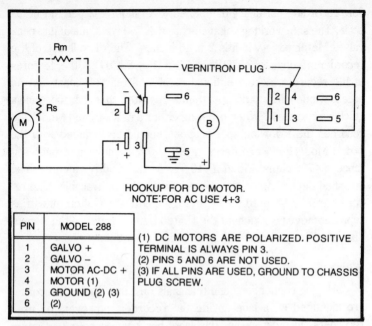

Fig. 4-2. Wiring diagram and terminal connections of the Rustrak Model 288 strip chart recorder.

1. Obtain a fresh roll of chart paper *24*.
2. Unscrew recorder thumbscrew *7*.
3. Unlatch retaining clips *17* and *25*.
4. Open chassis latch *54* located on the right front side of the recorder.
5. Remove supply roller *22* and take up roller *21*.
6. Slide supply roller *22* into full roll of chart paper *24*, with roller shoulder nearest paper perforations. Unroll about a foot and slide paper (back side up) between side plate and latch. Steer paper against drum *50* to clear pointer on meter assembly *52*. Engage paper perforations into drum sprockets and drop roll into seating notches.
7. Slide cardboard sleeve *23* onto take up roller *21*. Butt paper against disc and tape end of chart paper to sleeve. Turn one revolution for proper paper alignment.
8. Roll paper tightly and straight on take up roller. Keep paper taut. Drop into deeper notch, engaging gear *17*, and close gear retaining clips. Snap up chassis latch *54*. Close recorder by screwing in thumbscrew *7*. Advance paper by

depressing and turning chart advance wheel *6* located on front panel.

Tear Off Method (recommended). As with the previous procedure, numbers in italics are related to Fig. 4-3.

1. Obtain either a partial or full roll of chart paper.
2. Snap out nameplate *4* using a screwdriver.
3. Remove the two drive belts (item *2*) and replace nameplate.
4. Open recorder. Unlatch retaining clips *17* and *25*. Open chassis latch *54*. Remove supply roller *22* and take up roller *21*. Unscrew release screw two or three turns. Spring side plate *48*, and remove guide roller *28*. Slip roller through the two belts. Reseat roller. Tighten release screw.
5. Pass take up roller through belts. Seat the roller to engage drive gear. Pull the belts into chamfered center groove of guide roller *10* and align into the V slots on take up roller *21*. Insert supply roller *22* into spool of chart paper *24* (may be a partial roll), so that the roller shoulder is nearest the perforations. Unroll the chart paper and slide (back side up) between the drum sprockets and drop the roll into the seating notches.
6. Close the retaining clips. Snap up the chassis latch. Pull the drive belts into the V slots on the guide roller. Close the recorder.
7. Advance the paper and set the time by depressing and turning chart advance wheel *6*. Tear off paper as needed.

Chart Paper

The chart paper used in the Model 288 recorder is pressure sensitive and requires no heat or ink to produce markings. The paper has a black paper base coated with a thin, light-reflecting white layer. In use, the recorder stylus pushes aside the top coat, exposing the black-based paper. No buildup of coating material occurs on the paper or the stylus. The chart paper is easily handled and resists smudging. Chart spools are 63 feet long and permit continuous recording for 31 days when used at a chart speed of 1 inch per hour. Each foot of chart paper is numbered one through twelve to aid time setting. Each inch, in turn, is graphically divided into four equal

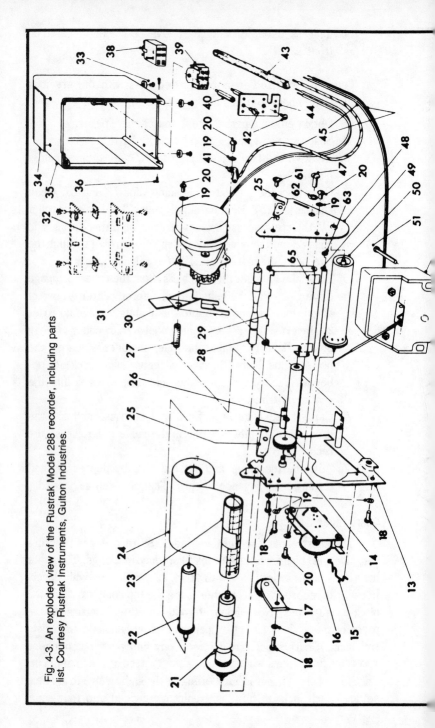

Fig. 4-3. An exploded view of the Rustrak Model 288 recorder, including parts list. Courtesy Rustrak Instruments, Gulton Industries.

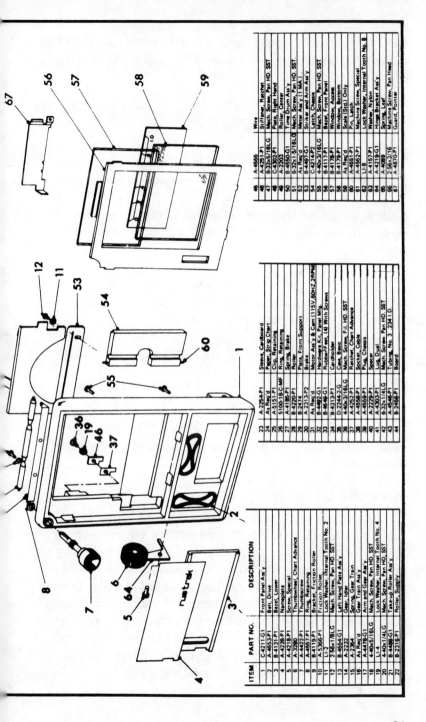

ITEM	PART NO.	DESCRIPTION
1	C-4211-G1	Front Panel As'y
2	A-4653-P1	Belt, Drive
3	B-4131-P1	Bezel, Lower
4	A-4228-P1	Nameplate
5	A-4218-P1	Screw, Special
6	A-3280	Thumbwheel, Chart Advance
7	A-4652-P1	Thumbscrew
8	B-4119-P1	Ring, Retaining
9	B-4119-P1	Bracket, Friction Roller
10	A-5366-P1	Friction Roller
11	Li-2	Lock Washer, Internal Tooth No 2
12	2-56x1/8LG	Mech. Screw, Pan HD SST
13	B-4554-G1	Left Hand Plate As'y
14	A-2232	Gear, Idler
15	A-2364	Spring, Gear Train
16	As Req'd	Gear Train As'y
17	A-4478-G1	Arm and Gear As'y
18	4-40x5/16LG	Mech. Screw, Pan HD SST
19	L1-4	Lock Washer, Internal Tooth No. 4
20	4-40x1/4LG	Mech. Screw, Pan HD SST
21	B-4498-G1	Take-up Roller As'y
22	B-2316-P1	Roller, Supply

ITEM	PART NO.	DESCRIPTION
23	A-2254-P1	Sleeve, Cardboard
24	As Req'd	Paper, Strip Chart
25	A-5151-P1	Clip, Retaining
26	5100.18-C-MF	Ring, Retaining
27	A-4185-P1	Spring, Brake
28	B-2474	Roller, Front Support
29	A-2313-P2	Plate, Front Support
30	B-2474	Brake
31	As Req'd	Motor As'y & Cam (115V, 60HZ, 2RPM)
32	B-4492-G1	Hardware Kit, Panel Mtg.
33	B-4549-G1	Rubber Feet, (4) With Screws
34	A-4313-P1	Cardholder
35	D-2244-2-G	Case, Basic
36	4-40x3/16LG	Mech. Screw, Fil. HD. SST
37	A-4252-P1	Ratchet, Chart Advance
38	A-4558-P1	Socket, Cable
39	A-4559-P1	Plug, Chassis
40	A-4277	Spring, Dual
41	A-4303-P1	Mech. Screw, Pan HD
42	6-32x1/4LG	Mech. Screw, Pan HD SST
43	A-4546-P1	Tubing, No. 3 .234 I.D.
44	B-2466-P1	Board

ITEM	PART NO.	DESCRIPTION
45	A-4555	Wire
46	A-4261-P1	Stiffener, Ratchet
47	8-32x5/16LG	Mech. Screw, Pan HD SST
48	C-4302-P1	Plate, Right Hand
49	A-2203	Roller, Center
50	B-4550-G1	Time Drum As'y
51	6-32x15/16LG	Mech. Screw, Pan HD SST
52	As Req'd	Motor As'y (MA)
53	B-4497-G1	Striker and Arm As'y
54	C-4214-P2	Latch, Chassis
55	4-40x3/16LG	Mech. Screw, Pan HD SST
56	B-4117-P1	Bezel, Front Panel
57	B-4178-P1	Window, Access
58	B-4179-P1	Window, Bottom
59	As Req'd	Pin, Latch
60	A-4651-P1	Machine Screw, Special
61	A-4552-P1	Lock Washer, Internal Tooth No. 8
62	A-4157-P1	Washer, Nylon
63	A-4219-G1	Hinge Bracket As'y
64	A-4099	Spring, Leaf
65	2-56x3/16	Machine Screw, Pan Head
67	A-4870-P1	Guard, Pointer

segments. A printed "end of roll" warning begins three feet from the end of the roll, and is repeated each foot to the end of the roll. The duration of a roll of chart paper can be easily computed by using the formula:

$$\frac{756}{\text{Chart speed (in./hr.)}} = \text{hours of operation.}$$

Notes, remarks, etc., may be written in pen or pencil on the chart paper while the recorder is operating. Even a paper clip or any blunt stylus can be used to make notations through the open access window. Many chart styles are available to suit the experimenter's needs. The following chart styles can be ordered either from Rustrak or Graphic Controls.

Style G or WG - 20 divisions
Style 1 or W1 - 30 divisions
Style H or WH - 40 divisions
*Style A or WA - 50 divisions

Style K or WK - 60 divisions
Style L or WL - 80 divisions

*Generally used for flare recording

The paper can be ordered, as mentioned before, from either Rustrak or Graphic Controls. Refer to Appendix B for the vendor's addresses.

Maintenance and Calibration

Although operation of the recorder is relatively trouble free, observing a few hints on care and maintenance will give years of trouble free operation. Do not attempt to oil any of the dry running gears in the recorder. Oil is the worst culprit; it leaves a sticky residue to which dust clings causing jammed gears and rollers. Clean the rollers and gears with a fine hairbrush, and lint free cloth. Many problems during use are due to improper chart loading, incorrect wiring, or mishandling. Remember that it is a precision instrument and should be given the best of care. Typical problems that may occur are shown in Table 4-1. When testing always refer to the exploded view in Fig. 4-3.

Table 4-1. Rustrak Model 288 Chart Recorder Troubleshooting

SYMPTOM	POSSIBLE CAUSE	SERVICE HINT
Meter records above or below zero with no signal applied.	Meter zero out of adjustment.	Remove front name-plate adjust mechanical zero.
Meter can't be zeroed.	Stylus bent; cross arm bent.	Straighten stylus or cross arm.
Meter reads zero with signal applied.	Meter open. Connections to meter open. Multiplier resistor open.	Check meter (Do not use VOM) Check wiring to galvo. Check plug and terminal wiring. Check multiplier resistor with VOM.
Meter hangs above zero.	Dust in meter pivots.	Carefully blow air into meter pivots (use rubber bulb).
Meter reads consistently below zero.	Polarity to meter reversed.	Check wiring to mating socket. Check polarity of signal being recorded.
Reroll mode: paper crinkles at view window	Take up spool not seated to engage drive gear.	Unlatch retaining clip reseat take up spool in further notch.
Reroll: paper doesn't drive through recorder.	Perforations not engaged in drive drum. Gear train not engaged to drive drum.	Reload paper. Bend tabs on gear train slightly for end play. Bend gear train spring for more tension.
Tear off mode: paper does not drive through tear-off slot.	Drive belts not seated on take up roller. Also riding in large center groove of drive belt roller.	Seat drive belts as explained in loading instructions.
Paper tears on drive drum.	Roll of chart paper has spiraled.	Reload, taking care chart paper is perfectly aligned.
Chart advance thumb-wheel is locked.	Tab that disengages gear is bent.	Remove gear train and straighten tab.

Mechanical Zero

If the trace on the chart paper is above or below zero with the galvanometer terminals shorted and the motor running, snap out the front nameplate (using a screwdriver in the slot at the left) and vary the zero adjust screw. The stylus also may be adjusted mechanically at any point upscale if the recorder is used over a narrow span of the calibrated scale. Calibrate against a standard (0–100 μA) meter across the *galvo* terminals and a source.

Changing Gear Boxes. Open the recorder. Remove gear box spring *15* shown on the exploded view in Fig. 4-3. Move the gear box in the direction of arrow on its case. Lift out the gear box from the *top*; do not attempt to force or lift it from the bottom. Insert the new gear box *bottom* in first. Slide in position against arrow direction. Replace the gear box spring. In some cases the experimenter may find that the 1 inch per hour chart speed is not suitable for the recording requirements. Figure 4-5 is a complete listing of gear train ratios, chart, speed, writing speed, and trace densities, suitable for any experimental requirement. The motors and gear train drives are stocked as standard items by Rustrak.

Formulas. Writing speed of the recorders varies with the motor speed. Chart speed and trace density depend on the ratio of the interchangeable gear box which couples the paper drive to the motor. By using the following simple formulas, any number of chart speed and writing speeds can be calculated.

$$ws = \frac{ms}{4} \qquad cs = \frac{ms \times gn}{2}$$

$$td = \frac{756}{cs}$$

Where:

ws = Writing speed in strikes/second
ms = Motor speed in rpm
cs = Chart speed in inch/hour
gn = Gear train assembly No.
td = Trace density in strokes/inch

Cleaning the Recorder. The coating used on the Rustrak recorders is Nextrel Brand Suede Coating by 3M. It was chosen because of its toughness and visual qualities. It can be cleaned with a damp sponge. The usual household spray and liquid cleaners can be used on tough stains without damage to the coating.

THE SIMPSON MODEL 2750 RECORDER

The Simpson Model 2750 is a clamp type, low speed strip chart recorder, shown in Fig. 4-4. It is manufactured by Simpson Electric Company of Elgin, Illinois. It too uses pressure sensitive paper and incorporates a rugged taut band meter movement as the measuring

Table 4-2. Rustrak Model 288 Chart Recorder Gear Train Ratios, Chart Speed, Writing Speed, and Trace Density.

DRIVE MOTOR (rpm)	WRITING SPEED	Parameter	60 (1:1)	45 (1.5:1)	30 (2:1)	15 (4:1)	12 (5:1)	10 (6:1)	6 (10:1)	4 (15:1)	3 (20:1)	2 (30:1)	1 (60:1)	1/2 (120:1)	1/4 (240:1)	1/8 (480:1)
1/2	1 strike/ 8 seconds	Chart Speed — English Units/Hour	15 inches	11¼ inches	7½ inches	3.75 inches	3 inches	2½ inches	1½ inches	1 inch	3/4 inch	1/2 inch	1/4 inch	1/8 inch	1/16 inch	1/32 inch
		Chart Speed — Metric Units/Hour	38 cm	28.6 cm	19 cm	95.3 mm	76.2 mm	63.5 mm	38 mm	25.4 mm	19 mm	12.7 mm	6.35 mm	3.2 mm	1.6 mm	0.8 mm
		Duration of Chart Paper Spool	50.4 hrs	67.2 hrs	100.8 hrs	201.6 hrs	252 hrs	302½ hrs.	21 days	1 month	6 weeks	9 weeks	18 weeks	9 months	72 weeks	144 weeks
1	1 strike/ 4 seconds	Chart Speed — English Units/Hour	2½ feet	22½ inches	15 inches	7½ inches	6 inches	5 inches	3 inches	2 inches	1½ inches	1 inch	1/2 inch	1/4 inch	1/8 inch	1/16 inch
		Chart Speed — Metric Units/Hour	76.2 cm	57.2 cm	38 cm	19 cm	15.2 cm	12.7 cm	76.2 mm	50.8 mm	38 mm	25.4 mm	12.7 mm	6.35 mm	3.2 mm	1.6 mm
		Duration of Chart Paper Spool	25.2 hrs	33.6 hrs	50.4 hrs	100.8 hrs	126 hrs	151.2 hrs	252 hrs	378 hrs	21 days	1 month	9 weeks	18 weeks	9 months	72 weeks
2	1 strike/ 2 seconds	Chart Speed — English Units/Hour	5 feet	45 inches	2½ feet	15 inches	1 foot	10 inches	6 inches	4 inches	3 inches	2 inches	1 inch	1/2 inch	1/4 inch	1/8 inch
		Chart Speed — Metric Units/Hour	1.52 mtrs	1.14 mtrs	76.2 cm	38 cm	30.5 cm	25.4 cm	15.2 cm	10.2 cm	76.2 mm	50.8 mm	25.4 mm	12.7 mm	6.35 mm	3.2 mm
		Duration of Chart Paper Spool	12.6 hrs	16.8 hrs	25.2 hrs	50.4 hrs	63 hrs	75.6 hrs	126 hrs	189 hrs	252 hrs	378 hrs	1 month	9 weeks	18 weeks	9 months
4	1 strike/ second	Chart Speed — English Units/Hour	10 feet	7½ feet	5 feet	2½ feet	2 feet	20 inches	1 foot	8 inches	6 inches	4 inches	2 inches	1 inch	1/2 inch	1/4 inch
		Chart Speed — Metric Units/Hour	3.05 mtrs	2.29 mtrs	1.52 mtrs	76.2 cm	61 cm	50.8 cm	30.5 cm	20.3 cm	15.2 cm	10.2 cm	50.8 mm	25.4 mm	12.7 mm	6.35 mm
		Duration of Chart Paper Spool	6.3 hrs	8.4 hrs	12.6 hrs	25.2 hrs	31½ hrs	37.8 hrs	63 hrs	94½ hrs	126 hrs	189 hrs	378 hrs	1 month	9 weeks	18 weeks
6	3 strikes/ 2 seconds	Chart Speed — English Units/Hour	15 feet	135 inches	7½ feet	45 inches	3 feet	2½ feet	18 inches	1 foot	9 inches	6 inches	3 inches	1½ inches	3/4 inch	3/8 inch
		Chart Speed — Metric Units/Hour	4.57 mtrs	3.43 mtrs	2.29 mtrs	1.14 mtrs	91.4 cm	76.2 cm	45.7 cm	30.5 cm	22.9 cm	15.2 cm	76.2 mm	38 mm	19 mm	9.53 mm
		Duration of Chart Paper Spool	4.2 hrs	5.6 hrs	8.4 hrs	16.8 hrs	21 hrs	25.2 hrs	42 hrs	63 hrs	84 hrs	126 hrs	252 hrs	21 days	6 weeks	3 months
8	2 strikes/ second	Chart Speed — English Units/Hour	20 feet	15 feet	10 feet	5 feet	4 feet	40 inches	2 feet	16 inches	1 foot	8 inches	4 inches	2 inches	1 inch	1/2 inch
		Chart Speed — Metric Units/Hour	6.1 mtrs	4.57 mtrs	3.05 mtrs	1.52 mtrs	1.22 mtrs	1.02 mtrs	61 cm	40.6 cm	30.5 cm	20.3 cm	10.2 cm	50.8 mm	25.4 mm	12.7 mm
		Duration of Chart Paper Spool	3.15 hrs	4.2 hrs	6.3 hrs	12.6 hrs	15.75 hrs	18.9 hrs	31½ hrs	47¼ hrs	63 hrs	94½ hrs	189 hrs	378 hrs	1 month	9 weeks
10	5 strikes/ 2 seconds	Chart Speed — English Units/Hour	25 feet	225 inches	12½ feet	75 inches	5 feet	50 inches	2½ feet	20 inches	15 inches	10 inches	5 inches	2½ inches	1¼ inches	5/8 inch
		Chart Speed — Metric Units/Hour	7.62 mtrs	5.72 mtrs	3.8 mtrs	1.91 mtrs	1.52 mtrs	1.27 mtrs	76.2 cm	50.8 cm	38 cm	25.4 cm	12.7 cm	63.5 mm	31.8 mm	15.9 mm
		Duration of Chart Paper Spool	2.5 hrs	3.4 hrs	5 hrs	10 hrs	12.6 hrs	15 hrs	25.2 hrs	38 hrs	50½ hrs	75½ hrs	151 hrs	302 hrs	604 hrs	50 days
12	3 strikes/ second	Chart Speed — English Units/Hour	30 feet	22½ feet	15 feet	7½ feet	6 feet	5 feet	3 feet	2 feet	1½ feet	1 foot	6 inches	3 inches	1½ inches	3/4 inch
		Chart Speed — Metric Units/Hour	9.14 mtrs	6.86 mtrs	4.57 mtrs	2.29 mtrs	1.83 mtrs	1.52 mtrs	91.4 cm	61 cm	45.7 cm	30.5 cm	15.2 cm	76.2 mm	38 mm	19 mm
		Duration of Chart Paper Spool	2.1 hrs	2.8 hrs	4.2 hrs	8.4 hrs	10½ hrs	12.6 hrs	21 hrs	31½ hrs	42 hrs	63 hrs	126 hrs	252 hrs	21 days	6 weeks
16	4 strikes/ second	Chart Speed — English Units/Hour	40 feet	30 feet	20 feet	10 feet	8 feet	80 inches	4 feet	32 inches	2 feet	16 inches	8 inches	4 inches	2 inches	1 inch
		Chart Speed — Metric Units/Hour	12.2 mtrs	9.14 mtrs	6.1 mtrs	3.05 mtrs	2.44 mtrs	2 mtrs	1.22 mtrs	81.3 cm	61 cm	40.6 cm	20.3 cm	10.2 cm	50.8 mm	25.4 mm
		Duration of Chart Paper Spool	95 mins	2.1 hrs	3.15 hrs	6.3 hrs	7.8 hrs	9.45 hrs	15.75 hrs	23.7 hrs	31½ hrs	47¼ hrs	94½ hrs	189 hrs	378 hrs	1 month
TRACE DENSITY		STRIKES PER INCH (DENSITY)	30	40	60	120	150	180	300	450	600	900	1800	3600	7200	14400

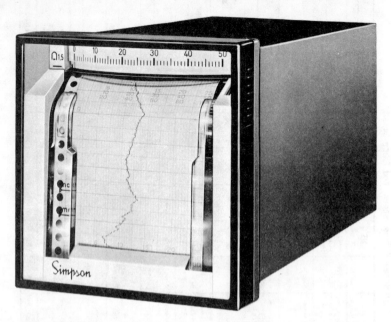

Fig. 4-4. The Simpson Model 2750 strip chart recorder. Photo courtesy Simpson Electric Co.

means. Indication accuracy is p/m 1.5 percent seconds. Response time of the meter movement is less than two seconds. The 2750 provides a recording by making a series of sequential impressions on pressure senstive chart paper. Every three seconds (six seconds on 10, 25, and 50 μA units) a metal bar clamps the pointer of the meter movement momentarily against the pressure sensitive paper. This, in turn, compresses the wax film on the paper and produces a black dot. Successive dots produce an apparently continuous line except at the highest chart speeds where the individual dots can be distinguished.

The meter movement is free to follow changes in the input value at all times other than during the momentary clamping period. At all other times a cam lifts the clamping bar above the pointer. The clamping bar and the chart drive are operated by different types of gear units available from the manufacturer. The standard gear unit supplied with the 2750 has a speed of 20/120 mm/hr. Three high speed gear units are available: 30/180 mm/hour, 60/360 mm/hour, and 100/600 mm/hour. When ordering the 2750, either new or used, the following specifications should be requested from the supplier: Model 2750 strip chart recorder; range 0– 100 μA motor drive 110 Vac. These recorders usually come with ac drive motors, although 12 Vdc drive motors can be special ordered from the factory.

Specifications

The meter movement, with taut band suspension, has a response time of less than 2 seconds. For ac measurements a full wave germanium diode bridge is used. The frequency influence is so designed that no additional error from 15 to 1000 Hz, 2.5 percent of full scale, from 10,000 to 20,000 Hz is tolerated. The dielectric test voltage is rated at 2000 Vac. Recording system specifications are as follows:

- Recorder. The chart paper and clamping bar are driven by a self contained synchronous motor.
- Motor Drive. A self starting 120V, 60 Hz synchronous motor is contained in the recorder. A three wire grounded line cord is also provided.
- Chart Speed. With the standard gear unit chart speed is 20 mm/hour (20 mm = 0.79 inch). This speed may be changed to 120 mm/hour by moving the gear change lever located on the right hand side of the chart drive unit. The gear change lever is used to provide a 6:1 speed selection with any gear unit.
- Chart Paper. The type of paper used is of the pressure sensitive type. The chart width is 2.3 inches and the length of the roll is 50 feet. Running time is 32 days. The chart paper varies according to the gear unit used and the number of scale divisions suitable for the recorded range. Table 4-3 can be used to select the proper chart paper catalogue numbers. The paper, usually supplied in quantities of 10 rolls per box, can be ordered directly from Simpson Electric or Graphic Controls. Consult Appendix B for addresses.

Operating Instructions

These instructions should provide the essential information required to set up and use the recorder. The numbers in italics refer to items located on Figs. 4-5 and 4-6.

Opening and Closing the Recorder. The Model 2750 recorder has a phenolic cover with a glass window. It is held to the case

Table 4-3. Simpson Model 2750 Chart Recorder Paper Options.

Divisions	Ranges	Chart Speed (mm/hour) vs Order No.			
		20/120	30/180	60/360	100/600
30	1.5-15-150 3-30-300 6-60-600	22310	22312	22314	22316
50	1-10-100 2.5-25-250 5-50-500	22311	22313	22315	22317

by a spring catch. To open, grip the fluted edges of the cover firmly, pull out at the top and lift slightly to disengage the cover from the case. When closing, be sure to place the two projections on the bottom of the cover into the corresponding holes in the case.

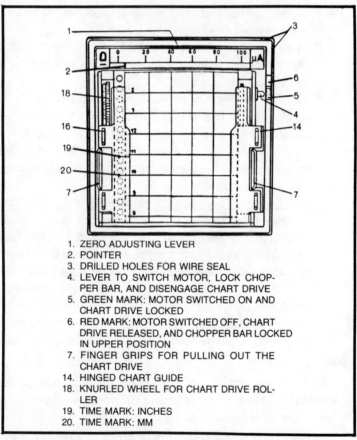

1. ZERO ADJUSTING LEVER
2. POINTER
3. DRILLED HOLES FOR WIRE SEAL
4. LEVER TO SWITCH MOTOR, LOCK CHOP-PER BAR, AND DISENGAGE CHART DRIVE
5. GREEN MARK: MOTOR SWITCHED ON AND CHART DRIVE LOCKED
6. RED MARK: MOTOR SWITCHED OFF, CHART DRIVE RELEASED, AND CHOPPER BAR LOCKED IN UPPER POSITION
7. FINGER GRIPS FOR PULLING OUT THE CHART DRIVE
14. HINGED CHART GUIDE
18. KNURLED WHEEL FOR CHART DRIVE ROL-LER
19. TIME MARK: INCHES
20. TIME MARK: MM

Fig. 4-5. Front view of the Simpson Model 2750 strip chart recorder including parts listing.

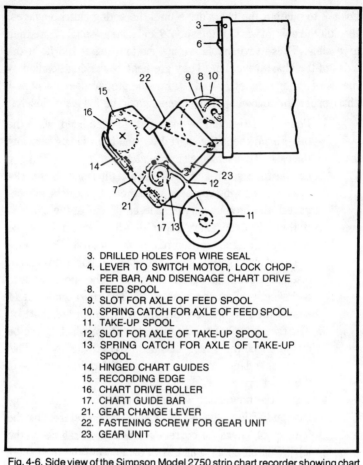

3. DRILLED HOLES FOR WIRE SEAL
4. LEVER TO SWITCH MOTOR, LOCK CHOP-
   PER BAR, AND DISENGAGE CHART DRIVE
8. FEED SPOOL
9. SLOT FOR AXLE OF FEED SPOOL
10. SPRING CATCH FOR AXLE OF FEED SPOOL
11. TAKE-UP SPOOL
12. SLOT FOR AXLE OF TAKE-UP SPOOL
13. SPRING CATCH FOR AXLE OF TAKE-UP
    SPOOL
14. HINGED CHART GUIDES
15. RECORDING EDGE
16. CHART DRIVE ROLLER
17. CHART GUIDE BAR
21. GEAR CHANGE LEVER
22. FASTENING SCREW FOR GEAR UNIT
23. GEAR UNIT

Fig. 4-6. Side view of the Simpson Model 2750 strip chart recorder showing chart drive pulled out and tilted.

Locking Lever. With the cover removed, locking lever 4 will be exposed in the upper right hand corner of the case. When the locking lever is pushed down in line with green section 5 on the label, the motor is switched on and the chart drive unit is locked in the case. When the lever is pushed up in line with the red section, the chart drive is unlocked, the clamping bar is locked into its upper position, and the motor is switched off. At this time the pointer of the movement is released and the chart drive can be pulled out of the case for replacing the chart roll or changing the chart speed.

Loading the Recorder. After having removed the phenolic cover, place the locking lever in line with red mark 6 and use finger holds *7* to pull out the chart drive until the spring clutch engages; then tilt the unit forward. Feed spool *8* and take up spool *11* are then accessible. These two spools are inserted in slots *9* and *12* at the sides of the chart drive unit. They are kept in working position by means of spring catches *10* and *13*. For the proper insertion of new chart paper the following instructions should be followed closely:

1. Feed spool *8* should be taken out with one hand, while the other hand keeps the chart drive unit from sliding back into the case. To disengage spring catches at *10*, get a firm grip on the feed spool and press at a right angle to the slot against the spring catches. With the spring catches disengaged, the feed spool will slide easily out of the slots.

2. If there is used chart paper on the spool, it must be removed before insertion of a new chart roll. To remove the used chart paper, grasp the take-up spool from below, disengage the spring catch by pressing the spool forward, and slide it out of the slots at *12*. The chart guides at *14* should then be opened to remove the recorded chart.

3. The take-up spool has a removable flange which has to be pulled off before sliding the used roll off the spool. This removable flange is located close to the right hand side of the chart drive unit.

4. Install the new chart with the perforations on the flange side and slide it against the flange. When reinserting the feed spool, check for correct position (perforations on the left). Pull off about 8 inches of paper from the feed spool and cut the end to a tapered point. With the chart guides open, insert the chart so that the pins of the chart drive roller engage the perforations of the paper. After replacement of the removable flange, the pointed end of the chart is inserted into the slot on the take-up spool, with the front side of the chart facing the inside of the spool and the perforated side of the chart strip touching the flange with the pinion gear. Wind a few turns on the spool to prevent the chart from disengaging. Close the chart guides, tilt the chart drive unit up and slide it into the case. Push the locking lever down in line with the green section, and replace the cover. When reinserting the take-up spool, make sure that the removable flange is on the right hand

side of the chart drive. Tighten the chart by turning the knurled flange on the feed spool.

Adjustment and Setting of Time Scale. On the Model 2750, recording is done at the clamping bar. Since this point is not visible from the front of the recorder, it is not possible to conveniently align the appropriate time line with the recording point. For this reason a reference point, a red line marked *mm*, is provided on the left chart guide. The distance between the recording point and the reference line is 60 mm. For various chart speeds, this distance will represent different time intervals. To determine the proper time interval adjustment, the cart speed must be known. The unit of the time scale on the chart depends upon the position of the gear change lever.

When the gear change lever is in the position marked ×6 the speed marked in parenthesis on the gear unit applies. For the standard gear unit, 20/120 mm/hour and the optional gear units 30/180 mm/hour, 60/360 mm/hour in the position marked ×6, each line represents 10 minutes. Since the chart has the lines numbered from 1 to 24, the chart can be read like a clock when the standard gear unit is used with the gear change lever in the ×1 position. In the ×6 position, the time is in multiples of 10 minutes. Thus a total of 240 minutes (4 hours) would be represented by a recording from one line marked 24 to the next one. On the chart for the 100/600 mm/hour gear unit, the unit of the scale is 6 minutes or 1 minute according to the gear change lever position. The scale is marked from 0 to 9, the numbers indicating 1/10 hours or minutes for the two different chart speeds.

For a given chart speed, the time adjustments for starting the recorder can be computed using the following formula:

$$\text{Time adjustment} = \frac{\text{Distance (60 mm)}}{\text{Chart speed}}$$

This time adjustment is then subtracted from the actual starting time and the chart is aligned at the *mm* mark to indicate the starting time minus the time adjustment. For example: assume that a standard gear unit was set up with the gear change lever in the ×1 position. The time adjustment from the chart would be 3 hours. If the actual time the recorder started was 10 o'clock, the *7* line would be aligned with the *mm* mark on the left chart guide.

Zero Adjustment on Chart. Disconnect the leads to the measuring circuit terminals and adjust the pointer with the zero adjuster (Fig. 4-5 Item *1*) until the recorded line coincides with the zero line on the chart.

Gear Unit Exchange. To remove the gear unit from the chart drive, loosen screw *22* in Fig. 4-6, slide the gear unit up, and tilt it out at the bottom to disengage the gears. To install the gear unit, engage the slots over the corresponding pins on the chart drive, making sure that the gears properly engage before locking the gear unit with screw *22*.

Selection of Chart Speed. For a specific chart speed, select the gear unit with the indicated speed on the label. If the speed desired is marked in parenthesis on the label, place the gear change lever in the position marked ×6. The other speed applies with the gear change lever in the position marked ×1. To change the position of the gear, change the depress lever to the release position.

Connections. Connect the measuring leads to the two instrument terminals mounted on the back plate of the recorder (see Fig. 4-7). To install the special line plug (which may not be furnished with a used recorder) to the recessed line terminals (marked 60 Hz, 100-127V), remove the screws from the two protruding studs with a screwdriver. Insert the line plug and replace the screws and washers to hold the plug in place. The third lead of the line cord is connected to the grounding screw. To do this, remove the grounding screw, place the washer and terminal lug of the grounding lead over it and reassemble, making sure that a good connection is made to the grounding terminal of the case.

Recording with Free Chart Exit (Tear Off Mode). For recording without a take up reel, an accessory clear molded acrylic cover is available from Simpson Electric. A snap-out section is provided in the bottom of the cover to allow for free exit of the chart. The chart paper may then easily be turned off and saved for a permanent record. When ordering this accessory refer to part No. 22395.

MAINTENANCE

The maintenance requirements for the Simpson Model 2750 recorder are basically the same as those of the Rustrak 288, with one exception. Every 1000 hours, lubrication of the drive bearings with

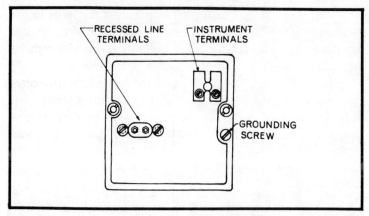

Fig. 4-7. Wiring connections on the back plate of the Simpson Model 2750 strip chart recorder.

light oil such as 3 in 1 or sewing machine oil is recommended. Deposits of dust on the scale and the clamping bar should be removed with a soft brush. The troubleshooting table 4-1 can also be applied to the Model 2750 since they both use the same basic drive system and paper take-up, with the exception of the belts in the tear off mode as used by the 288. The 2750 uses gears throughout the system.

SELECTING USED RECORDERS

The recorders that we have just described are the recommended type to be used with the solar flare receivers described in Chapter 3, although voltage (millivolt type) recorders can also be used with some slight modifications. Surprising as it may seem, hundreds of Esterline Angus, Varian, Brush, and Honeywell recorders can be purchased used or acquired from some industrial laboratories for a small fee. With a small amount of work these voltage-type recorders can be put into operating order. However, a word of caution when acquiring a used recorder. It is advisable to stay away from the models that use special paper or elaborate drive systems. These models may look quite impressive, but the paper they use can be very hard to obtain. Also, the larger the paper size the more expense. A good rule of thumb to use is not to go over a 6 inch strip chart. Since these recorders have basically voltage (millivolt) inputs, and the outputs of the flare receivers are set up for current (μA), some sort of signal conditioning device is required to

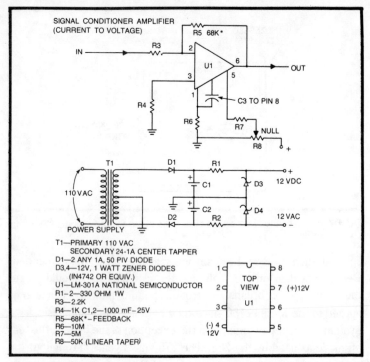

Fig. 4-8. Schematic diagram of a current-to-voltage signal conditioner amplifier.

boost the weak current signal to a voltage source to drive the recorder.

Figure 4-8 shows an inexpensive method of accomplishing this. By simply changing the values of R5, which is a feedback resistor, the gain can be changed. Increasing the value of the resistor increases the gain of the unit; decreasing the value of the resistor decreases the gain. The schematic also shows plans for a companion power supply for the signal conditioner, but any 12 volt filtered supply can be used. Used voltage and current recorders can be obtained from any of the vendors listed in the Appendix. Edlie Electronics of Levittown, Long Island, New York, publishes a catalogue of about 80 pages devoted to electronic gear. A large section of the catalogue is devoted to used test gear and recording systems. They attempt to keep a stock of recorders available to amateurs/experimenters who order them on a first come, first served basis; however, they also list numerous voltage type recorders that can be used with the signal conditioner amplifier.

104

Chapter 5
Putting It All Together

The solid state VLF receivers and associated recorder interfaces that we have described in Chapters 3 and 4 were designed for two principal reasons: to record solar flares as they occur, and to be able to assemble a solar flare station well within the modest means of the amateur/experimenter. This chapter will deal with some of the problems encountered while monitoring solar flares, along with a discussion of the methods used in predicting radio propagation. These flare receivers are to be used as *indicators* to monitor the level of solar activity and should not be misinterpreted as a method to forecast or predict radio propagation; however, it is important to remember that when sunspots (groups of sunspots create solar flares) are few and far between, radio communication suffers, but when sunspots are numerous and close together radio conditions are excellent.

ASSEMBLING THE SOLAR FLARE STATION

The solar flare station should be housed in an area that is relatively free from manmade interference. The largest operational problem associated with the LF and VLF sudden enhancement of atmospherics (SEA) monitors is that of distinguishing solar flares from manmade electrical noise and interference. Manmade sources of the LF and VLF radio noise nearly always come from some type of spark-discharge apparatus, such as spark plugs, types of fluorescent lights, oil furnace ignitors, and motors using electrical brushes, such as electric drills, hair dryers, and mixers. With some experience, the amateur/experimenter can distinguish between a valid SEA, and madmade noise simply by observing the quality and time scale effect of the recorded signal. If the signal appears to be intermittent (switching on and off at a regular rate), chances are that it is of man-made origin. The sudden enhancement of signal (SES) monitors are less prone to this type of noise since they constantly receive a transmitted signal, which is usually stronger than the manmade. interference. If the station experiences a noise problem that cannot be controlled the use of some type of shielding is strongly recom-mended. Aluminum foil and pure tin have excellent shielding characteristics. By simply making a small box enclosure out of either of these metals and placing it over the input coils of the LF receiver, the noise problem usually can be reduced or eliminated. Another point to check is the type of wire used between the receiver output and the recorder input. Shielded wire here is also recommended since the small amount of current can easily be saturated with stray noise and end up as a questionable trace on the recorder.

A method that has proven itself in eliminating about 80 percent of a noise problem was based on an enclosure of wood (a box approximately 2 feet square). The top was left open and became the front of the enclosure. The bottom of the box and the sides were lined with tin plate, although aluminum foil could have been used. The power supplies were placed on top of the box, and the receiver and recorder were placed inside. The front part of the box was simply covered by a flap made of aluminum, but aluminum foil could have been used. Figure 5-1 shows the construciton of this type of enclosure. Heavy insulation type aluminum foil is recommended for the shielfing because of its rugged construciton and, if used as a shield in the enclosure, its ability to be easily stapled to the sides and back wall.

Improving Earth Ground Characteristics

As we mentioned earlier, the successful operation of the solar flare station depends a great deal on a good earth ground. The grounding system described in Chapter 3 that is used with the transistorized receiver is an ideal method of obtaining a good ground; however, the amateur/experimenter may live in an arid area such as

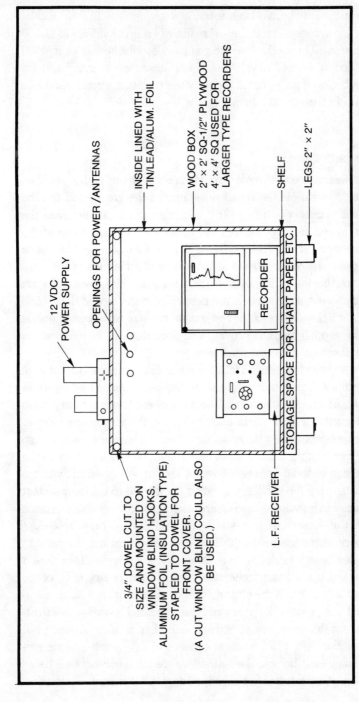

Fig. 5-1. Construction of a shielded enclosure used for solar flare recording.

12 VDC POWER SUPPLY

OPENINGS FOR POWER / ANTENNAS

INSIDE LINED WITH TIN/LEAD/ALUM. FOIL.

WOOD BOX
2′ × 2′ SQ-1/2″ PLYWOOD
4′ × 4′ SQ USED FOR LARGER TYPE RECORDERS

SHELF

LEGS 2″ × 2″

RECORDER

STORAGE SPACE FOR CHART PAPER ETC.

L.F. RECEIVER

3/4″ DOWEL CUT TO SIZE AND MOUNTED ON WINDOW BLIND HOOKS. ALUMINUM FOIL (INSULATION TYPE) STAPLED TO DOWEL FOR FRONT COVER. (A CUT WINDOW BLIND COULD ALSO BE USED.)

a desert where he may have an unusual soil problem. In areas such as this and in rocky terrain the soil resistivity is in the vicinity of 12,000 ohms/cm. The method we are going to describe here is a simplified method of *ground screen* techniques described in many antenna handbooks. If constructed properly the resultant ground resistance will bein the vicinity of one ohm or less.

The Earth Ground

Resistance to current through an earth ground path is based on three factors: (1) the actual resistance of the ground rod and the metallic connection to it; (2) the contact resistance between the ground rod and the adjacent earth; and (3) the resistance of the surrounding earth. The first factor should be almost negligible, since copper rod used for grounding purposes is of sufficient size and cross section that the total resistance value is small. The second factor can be a problem if the earth is not tamped properly around the ground rod, or if the surface of the rod is contaminated with grease, oxides or some other insulating cover. The third factor is the most critical and deserves the most consideration.

A ground rod driven into the earth radiates *some* current in all directions. It can be thought of as being surrounded by layers of earth, all of equal thickness. The layer of earth nearest the ground rod has the smallest surface area so it has the greatest resistance. The next layer has a larger surface area and less resistance and so on. Eventually a distance from the ground rod is reached where the inclusion of additional earth layers does not add significantly to the resistance of the earth surrounding the ground rod. Elaborate test equipment is available to measure analytically this optimum distance and the exact earth resistance, but the average amateur/ experimenter has neither this type of equipment nor the need to define these parameters to such an exacting degree. Therefore it seems that the basic problem is conductance, or lack of it, of the earth layers within the critical distance. Public utilities have recognized this problem for years and use a method known as *soil treatment* for dry sand, dry soils, rock and other problem ground (soil) conditions. The soil treatment consists of mixing salt into the surrounding earth layers. The method we are about to describe is based on this soil treatment but expanded somewhat to improve its effectiveness. As shown in Fig. 5-2, a hole 18 inches in diameter and 4 feet deep is dug in the soil. In the center of this hole drive a 3/4 inch × 8 foot copperweld ground rod into the earth as far as you can. A handy driving tool can be made from pipe that will barely pass over the rod. Take a threaded piece of pipe about 12 inches long and screw a coupling and hex head barstock plug onto the upper end. This is then placed over the ground rod to prevent mushrooming of the top while pounding on the pipe with a sledgehammer. After driving the ground rod in the hole, insert a plastic water pipe to one side and fill the hole with a chemical mixture containing the following mixture: *bentonite*, a clay-like soil commonly used by farmers in earthen dams and tanks to increase their ability to hold water without loss due to seepage (the material swells when wet and packs the mixture in the hole under adequate pressure to make up for poor tamping); *gypsum*, a popular product in the building industry which is used in the making of Sheetrock (this material gathers moisture and holds it in order to remain stable, and prevents loss of conductivity due to drying out); *rock salt*, the low resistance additive. The mixture, as homogeneous as possible, is tamped firmly into the hole to within 6 inches of the surface. Then a dirt cap is placed over the top of the mixture so that grass can grow over the spot where the hole was dug. The watering pipe is charged with a saturated solution

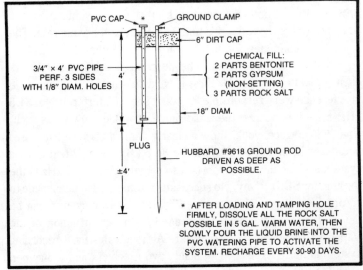

Fig. 5-2. Drawing of a method used for improving earth-ground characteristics.

of brine, which is recharged every 30 to 90 days. Over a period of years the salt leaches farther into the surrounding earth, increasing the effective inclusion of the surrounding earth layers. Connecting two of these chemical vats in parallel to form a ground system can reduce the combined resistance to a fraction of an ohm. Using this method, the amateur/experimenter can operate a solar flare station in virtually any arid or rocky terrain.

INTERPRETING RECORDED DATA

As we had previously discussed, valid SEA and SES type flare activity is hard to distinguish, and the experimenter can become proficient in distinguishing true SEA and SES receptions only by experience. In some cases this comes with a few months of recording; in other cases the experimenter may feel completely frustrated after hopelessly scanning miles of chart paper. Distinguishing solar flares from manmade interference and other types of radio noise can only come from experience; however, to make the job easier we have included valid SEA and SES charts that can be used for comparison purposes. The recording shown in Fig. 5-3 shows two traces taken on October 1, 1972, from the author's 34.5 and 75 kHz SES flare station. Looking at the top portion of the chart you will notice that the trace from the 75 kHz receiver follows relatively the same path as the trace from the 34.5 kHz receiver. At noon (EST) a large SES occurs and both receivers pick up the trace simultaneously. They both have the same rise time and decay at the same rate. The trace then continues for 2-1/2 hours and another small SES is recorded on both receivers. The purpose of using this dual frequency monitor is twofold. First it validates the SES. If manmade interference were responsible for some sort of interference at 34.5 kHz, since it is nearest the audio range, it would show up as such. The 75 kHz trace would show a clearly defined trace (a true SES) because it is further away from the 34.5 kHz noise. When using a system such as this it is important to remember that 75 kHz is the top limit for SES activity. No successful attempts have been made to record SES above 75 kHz. The 75 kHz receiver system can be designed around the schematics used in the construction of the Variable Frequency VLF Receiver. Although this flare recording system is more expensive to build (two receivers and a dual pen recorder) the results will justify the expense.

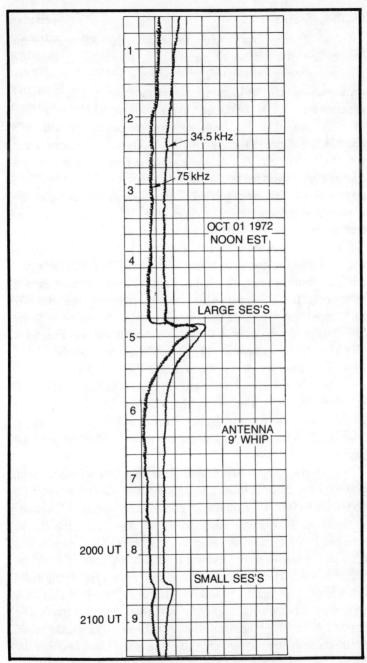

Fig. 5-3. Solar flare SES type trace taken on a dual pen recorder at 34.5 and 75.0 kHz, illustrating the comparison method of frequency recording.

111

RECORDING SEAs

The SEAs are not as easily recorded as the SES type of solar activity due to a number of reasons. The first being that we are attempting to record a natural phenomenon and have to take into consideration that this is pure static or radio noise. The amateur/experimenter should not attempt to record this type of solar activity as a first project; rather he should become acquainted with SES recordings and gradually ease into SEAs. The second reason is that at these low frequencies even the least bit of manmade interference may play havoc with the recording. Any attempt made to record SEA activity should first be preceded with an investigation of the experimenter's area where the recordings are to be made. A quiet interference free area is a must for SEA recordings. Figure 5-4 (Chart No. 1) shows the trace of an 18.6 kHz trace is very erratic and does not follow the fine line trace of the 34.5 kHz (SES) trace; however, the trend remains the same, that is, they both show definite signs of a flare occurring at sunset. Also you may notice that the 18.6 kHz SEA trace tends to drop off after the flare occurs, while the 34.5 kHz SES trace does not decay. If the trace were tuned to an LF signal at 18.6 kHz, the decay time would be much longer and the trace would follow the same pattern as the 34.5 kHz (SES) signal.

Chart No. 2 shows traces of the same receivers taken on February 27, 1973. Here again the SEA and SES activity travel basically the same chart path, but after the solar flare occurring at noon the 18.6 trace does not decay as fast as it did in the previous day's recording.

Chart No. 3 shows a recording of the same flare station taken on February 28, 1973. In looking at the chart you will notice that from 1600 UCT to 1900 UCT (universal coordinated time, or UT, universal time) the 34.5 kHz SES station was off the air, possibly due to scheduled maintenance or technical problems. At approximately 2200 UCT a flare occurred and was tracked by both the SEA and the SES receivers. The recording showed a very prominent hump in the enhancement of signal strength at 34.5 kHz. The 18.6 kHz SEA trace was increased but not nearly as much as the SES trace. The basic SEA and SES receiver described in Chapter 3 is ideally suited for this type of comparison trace. By simply switching to either the SEA or SES frequency for a period of hours, you can compare traces such as we have described and eliminate any outside manmade interference. The ideal system would have two of these SEA/SES receivers and monitor either SEA or SES with individual recorders, for purposes of comparison and validation of flare activity.

FREQUENCY CORRELATION PROBLEMS

The comparison method used for validating solar flares *does* work approximately 90 percent of the time, though for some mysterious reason this is not always true. As we discussed earlier, SES of transmissions at various frequencies does differ. By looking at the two charts in Fig. 5-5 this appears to be true. Looking at the top part of the chart you will see that the 18.6 kHz trace on February 18, 1973, shows a disturbance only at this frequency. The 34.5 kHz trace shows a prominent increase leading to a hump, gradual decay, then shoots straight up to full scale. This the author terms as a pre-flare disturbance—a rather rare type of indication. If this type of disturbance is recorded at 18.6 kHz, it may be safe to assume that a solar flare of large magnitude will be recorded on the SES receiver. Chart No. 2 shows again the same 18.6 kHz disturbance recorded on February 18, 1973, with the 34.5 kHz trace staying well below the limit of the 18.6 kHz trace. The traces then continue, with the 34.5 kHz system recording a small flare approximately 2 hours from the disturbance then gradually decaying before climbing off scale for an SES of great magnitude.

ANTENNA EFFECTS

As you may have noted, all of the charts used in the comparison study method were monitored using a 9 foot CB type whip antenna. The recommended antennas are the 108 inch CB whip for the first two receivers in Chapter 3 and the Maag downspout vertical antenna for the last receiver in that chapter. The long wire antenna, which has numerous uses at the higher end of the LF band, is almost useless with the solid state receivers described in Chapter 3. For comparison antenna traces, a 100 foot long wire and 9 foot whip antenna were attached to two receivers, one tuned to 34.5 kHz, the other tuned to 24 kHz (both SES). By observing the tracings in Fig. 5-6 you can see the difference between the long wire and the whip antenna. The 9 foot whip picks up all of the frequency variances, while the long wire shows a relatively straight line and gradually drops off in signal strength. These receivers were designed for short

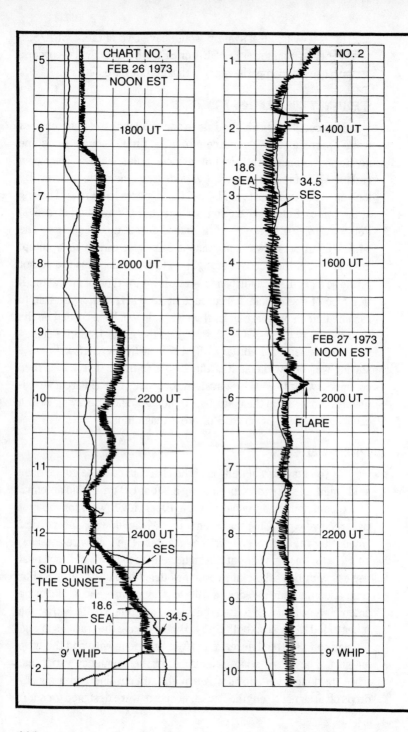

CHART NO. 1
FEB 26 1973
NOON EST

1800 UT

2000 UT

2200 UT

2400 UT
SES
SID DURING
THE SUNSET
18.6
SEA   34.5
9' WHIP

NO. 2

1400 UT

18.6
SEA   34.5
SES

1600 UT

FEB 27 1973
NOON EST

2000 UT
FLARE

2200 UT

9' WHIP

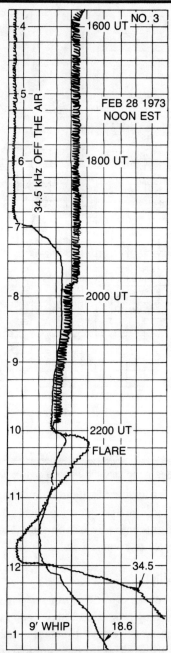

Fig. 5-4. A three day trace of solar flare recordings using 18.6 (SEA) and 34.5 (SES) kHz receivers for comparison purposes.

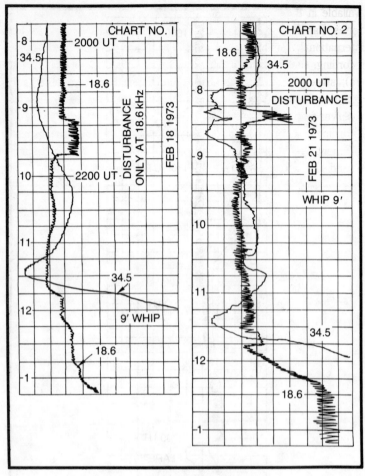

Fig. 5-5. A pre-flare disturbance at 18.6 kHz.

vertical antennas. If a long wire antenna is available it *can* be used; however, an impedance matching device should be employed. The short whip vertical antenna is by far the easiest to erect and gives traces that are simple to compare and analyze.

EFFECTS OF THE SUN

The sun is a source of radiation of radio frequencies. These emissions are caused by plasma oscillations and gyro-oscillations in the solar atmosphere, as well as by noise originating in random collisions of the electrons. The flux of solar radio noise at the earth's surface is monitored at a number of observatories, and the data is available in the Central Radio Propagation Laboratory (CRPL) F Series, Part B, and in the Quarterly Bulletin on Solar Activity of the International Astronomical Union. There are several methods of recording, such as fixed frequency (similar to our SES-SEA receivers), sweep frequency, total flux, and interferometers (used for exploring the distribution of radio sources on the solar disc). A list of observer stations, the frequencies used, and the universal times of recording are also listed in the Bulletin. The standard unit of measurement of flux is 10-22 watts per square meter per cycle per second.

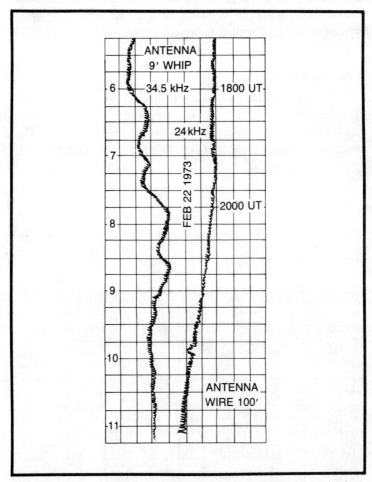

Fig. 5-6. Receiver trace patterns using a long wire antenna at 24 kHz and a short vertical at 34.5 kHz, showing characteristic differences.

Sunspots

One of the most notable phenomena on the sun's surface is the appearance and the disappearance of certain dark areas known as sunspots. The life of a sunspot is highly variable; some spots last only a few days, whereas a few may survive four or five solar rotations (of about 28 days each). Their exact nature is not known, but they appear to be vortices in the matter comprising the photosphere. Sunspots appear dark because the surface temperature is only about 3000°K, compared with the 6000°K of the quiet photosphere. Sunspots tend to group together. A group may contain a few isolated spots or dozens of spots. One of the more interesting features of sunspots is their unusually strong magnetic fields.

The Sunspot Cycle

The great variation of sunspot activity in 1954 (low) and in 1958 (high) can be seen in Fig. 5-7. Clearly there is a variation in the sunspot activity. To measure this activity an index is required. The most common index of sunspot activity is the Wolf number R given by:

$$R = k (10g + S)$$

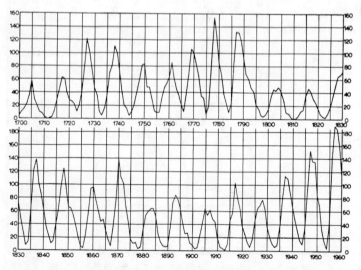

Fig. 5-7. Sunspot activity in the years 1700 to 1830 and 1830 to 1960.

Where

R = activity index

K = correction factor for equipment and observer characteristics

G = number of groups

S = number of observable individual spots

It can be seen that this number is weighted so heavily in favor of groups that its value as an index of sunspot activity is questionable. Nevertheless, it is valuable because of its availability for a period of about 200 years. It, therefore, provides a large homogeneous sample of data. Sunspot numbers are available on a daily basis and for monthly and yearly averages. When the sunspot numbers are plotted for a long period of time, such as shown in Fig. 5-7, it is clear that the sunspot activity has a periodicity of about 11 years. The duration of an individual cycle may vary over much wider limits; therefore, the periods between minima range from about 8.5 years to about 14 years and those between maxima range from 7.3 years to about 17 years. Furthermore, the annual minima of R lie in the range of 0 to 10, whereas the maxima range from about 50 to 190. The daily values of R vary between 0 and 355 or more. In addition to the 11 year period, there are periods, particularly near the 11 year minima, when a 28 day (approximate) recurrence is evident. This period corresponds to a solar rotation. Many attempts have been made to represent, analytically, the existing sunspot data with a view to the prediction of future cycles. These predictions have never been particularly successful; however, new methods have been tried and are still in the experimental stage. In addition to the relatively dark sunspots, bright areas often appear on the solar disc. These areas, called *plages*, are observed in the light of calcium K and are intimately associated with sunspots. Plages are comparatively long-lived active regions; they precede in appearance the spots, which develop in them, and persist after the spots have disappeared.

Solar Flares

A solar flare is a burst of light occurring in the chromosphere near a sunspot. It is most easily observed in the Hα at a wave length of 6563A, although in some rare cases flares have been seen in white light. At present, solar patrol observations are almost continuous, so most of the flares occurring in the solar hemisphere facing the earth are detected. Flares are frequent occurrences, *particularly at the peak of the sunspot cycle*. Flares are divided into classes of importance 1−, 1+, 2, 2+, 3 and 3+ according to area and brightness. The average duration of a flare lasts about 30 minutes, and three flares last about 60 minutes. Of course, the life of an individual flare may vary greatly from the mean value. Flares smaller than importance 1 are referred to as subflares. The development of a flare is somewhat as follows: a rapid rise (flash) to peak intensity, a brief period of peak intensity, and a steady decay or decline. There is a close statistical relation between the number of flares per solar rotation, Nf, and the corresponding mean sunspot number $\overline{R}$, given by $Nf = a\,(\overline{R} - 10)$. Therefore, the mean value of a for the sunspot cycles having peaks in 1937 and 1947 are 1.98 and 1.47, respectively. During some flares there is a marked increase in the flux of solar ionizing radiation in the far ultraviolet and soft x-ray region of the spectrum. This results in enhancements in the electron densities in the D and E regions as we had previously discussed in the preceding chapters.

Radio Emissions

The amplitude of the solar signals remains relatively constant for long periods of time, and then will be greatly enhanced during a noise storm. These storms are often associated with solar flares and certain geophysical disturbances, and may last for a period of hours or days. Noise bursts are categorized as follows:

• *Type I or Noise Storms* occur most frequently on frequencies less than about 300 MHz. Type I bursts extend over a narrow frequency range, 5 to 30 MHz, for instance, while their lifetimes range from 0.2 seconds to 1 minute or so. During the lifetime of a single burst, the frequency of emission may drift with time to higher or to lower frequencies. Certain outbursts, which last on the order of several minutes, are associated with the larger solar flares, and may involve signal enhancements of the order of a million times the quiet values.

• *Type II (Slow Drift) and Type III (Fast Drift)* bursts are the drift of the emission to lower frequency with time. This drift can be explained due to a source moving outward through the corona, emitting at the plasma frequency, which decreases as the electron density decreases. Type III events are very frequent.

•*Type IV* events are long-lived emissions emanating from a source of large extent that move to heights of the order of a solar radius. Radiation occurs over a broad band of frequencies at each height. The characteristics of Type IV radiation can be explained by the synchrotron emission of electrons. From the point of view of sun-earth relationships, Type IV events are particularly important because of the strong association with solar corpuscular radiation, evidenced by polar cap absorption events, ground-level cosmic ray increases, and great magnetic storms.

•*Type V* is continuum emission of shorter duration, following a Type III burst.

IONOSPHERIC PROPAGATION PREDICTIONS

Another aim of this chapter is to give the reader a brief idea of the way in which ionospheric data is applied to the problems of high frequency communication via the ionosphere, the problems concerned with the prediction of *optimum* working frequencies (the use of as high a frequency as possible to reduce ionospheric absorption). *System losses* are dealt with in *Theoretical Considerations for Selection of Optimum Frequencies* by the Central Radio Propagation Laboratory. The reader is urged to consult that work.

Purpose of Predictions

Because the ionosphere varies from hour to hour and from day to day, it is necessary to have a knowledge of this variability in order to select the optimum frequency, required transmitting power, antenna configuration, and so on. Broadly speaking, there are two types of predictions required.

1. Relatively short term frequency predictions are required by the radio operator in order that he may be able to anticipate maximum useful frequency (MUF) failure and thus increase communications reliability.

2. The long term predictions are required for the planning of station equipment. This involves the choice of antennas, operating frequencies, and power. The sort of questions they are faced with are: What will be the range of usable frequencies over the next sunspot cycle? What is the

minimum power necessary to fulfill the requirements of the circuit? What elevation angles must be used in antenna design? It might be noted here that yet another type of prediction, if it may be called a prediction, is short term information obtainable from oblique ionosodes, or back scatter observations. This type of prediction is of importance only to commercial traffic control centers and those concerned with circuit operation. For the first type of prediction, the most accurate method is to predict median values of the required parameters month by month. Such monthly predictions are issued usually about three months in advance and include observed ionospheric data up to one year before the prediction is made. The solar cycle information required is usually based on even more recent data. Predictions of type (2) are needed over a complete sunspot cycle and normally include data over the solar cycle prior to that in which the prediction is made. The behavior of the E and F1 layers of the ionosphere is so regular that permanent nomograms can be used for determining the required characteristics. On the other hand, the F2 layer is very irregular, and since it is the most important from the point of view of high frequency communications, maps have been prepared giving the variation of $f_0 F_2$ in addition to the M factor, as a function of time of day, season, and sunspot number. The M factor is the maximum usable frequency factor.

Predictable Characteristics Sunspot Number

The long term variation of ionospheric parameters is tied closely with the sunspot cycle, and although no *completely* satisfactory measure of solar activity is available, the 12 month running average sunspot number is most widely used. Normally the most practical way to determine the sunspot number is to use the predictions of others. When predictions of future cycles are required, an average minimum of 10 and an average maximum of 130 are recommended as a rule of thumb. New indices, based upon other parameters such as ionospheric noise and solar noise, are presently being investigated.

Maximum Frequencies

The predictions of $f_0 F_2$ are based on the seasonal, and geographical variations/or sunspot number dependence are established from previous sunspot activity. No attempt is made in these long term predictions to estimate day to day fluctuations in critical frequency, since these are rather localized in both space and time. The basic ionospheric data used in predictions are the E, F_1, and F_2 layer ordinary wave critical frequencies, and the M (3000) F_2, that is, the M factor for a distance of 3000 km. The behavior of the E and F_1 layers is such that critical frequencies can be predicted fairly accurately, in terms of solar zenith angle x, and $\overline{R}$ nomograms can be constructed to give f_0 E. The behavior of the F_2 layer is more complicated than that of the E and F_1 layers and cannot be represented analytically. In the preparation of F_2 predictions, the first step is to obtain the line of best fit of the monthly median values of f_0 F_2 to the 12 month running average sunspot number. This is done for each hour (or each alternate hour) of the day and for all stations. The slopes and intercepts of these lines are then used to construct worldwide maps from which, with a knowledge of $\overline{R}$, it is possible to calculate f_0 F_2. Sometimes, the value of f_0 F_2 at some specific $\overline{R}$ ($= 50$ for example) is given instead of the intercept or together with it to give the slope. The predictions of the $M(3000)F_2$ are prepared in a similar manner. In those prediction methods (which are published monthly) the sunspot, seasonal, and diurnal variations are automatically taken into account. Given the F_2 zero distance maximum frequency, MUF, (zero) F_2 (fx = F_2) and the 4000 km maximum frequency, it is possible to interpolate for intermediate distances by means of Fig. 5-8. Note that the $M(4000)F_2$ is obtained by multiplying $M(3000)F_2$ by 1.1.

Radiation Angle

This depends on the height of reflection (assumed specular) of the waves. In the case of the E layer, it is sufficiently accurate to take the angles given in Fig. 5-9. This figure can also be used to give rough values in the case of F 2 layer propagation but, in general, it is more accurate, if the height of reflection is known, to determine the elevation angle from Fig. 5-10. The monthly median heights of the F 2 layer are given on a worldwide basis published in the National Bureau of Standards Bulletin 63D.

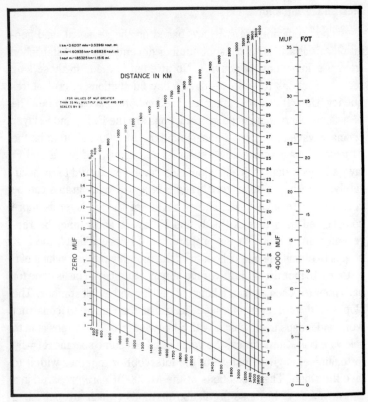

Fig. 5-8. Nomogram for transforming F_2 (zero) MUF and F_2 (4000) MUF to equivalent maximum usable frequencies for intermediate transmission distances. This is also a conversion scale for obtaining optimum working frequency (FOT). The nomogram for a given frequency from the MUF (zero) F_2. (From CRPL work sheets.)

FREQUENCY PREDICTION SYSTEMS

Data is also available from a variety of sources for short term and long term predictions in a form ideal for those with access to computer facilities.

Monthly Predictions

A variety of ionospheric predictions is issued monthly by laboratories in a number of different countries, usually three months or so in advance. Many of them are for localized areas, but several attempt worldwide predictions. While all prediction services use essentially the same ionospheric data and established physical relationships, there is considerable variation in the form of the presentation and in the approximations used in applying the various predictions to propagation problems. The CRPL prediction system, based on a proven method of numerical mapping of the ionospheric characteristics developed by Gallet and Jones using electronic computers, will be described briefly. The term *numerical map* is used to denote a function $\Gamma(\lambda, \theta, t)$ of the three variables: latitude (λ), longitude (θ), and time (t). The function $\Gamma(\lambda, \theta, t)$ is obtained by fitting certain polynomial series of functions of the three variables to the basic ionospheric data. The general form of $\Gamma(\lambda, \theta, t)$ is the Fourier series

$$\Gamma(\lambda,\theta,t) \pm a_0(\lambda,\theta) + \sum_{j=1}^{H} \left[a_j(\lambda,\theta)\cos jt + b_j(\lambda,\theta)\sin jt \right],$$

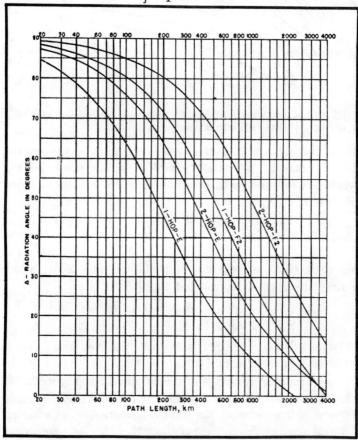

Fig. 5-9. Radiation angle versus path length, based on virtual reflection heights. The E layer is 105 km; the F2 layer is 320 km. (From CRPL work sheets.)

125

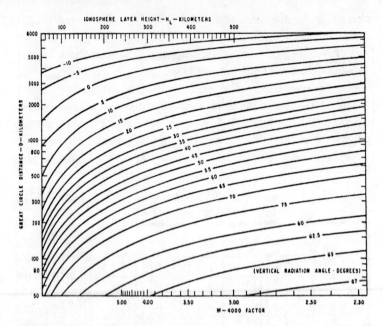

Fig. 5-10. Radiation angle as a function of great circle distance and ionospheric layer height. (From CRPL work sheets.)

Where H denotes the number of harmonics retained to represent the diurnal variation. The Fourier coefficients, aj $(\lambda, \mathbf{u})$ and by jb (l, θ), which vary with geographic coordinates, are represented by series of the form— k

$$\sum_{k=0}^{} Dsk\ Gk\ (\lambda,\ \theta)$$

Where the Gk $(\lambda,\ \theta)$ are given in Table 5-1 the index denotes which Fourier coefficient is represented, in the order given by

$$S = 2j,\ \text{For Aj}\ (\lambda,\ \theta),\ j = 0,\ 1,\ ...H,\ (a)$$
$$S = 2j\text{-}1\ \text{For bj}\ (\lambda,\ \theta),\ j = 1,\ 2,\ ...H,\ (b)$$

A numerical mpa, $\Gamma\ (\lambda,\ \theta,\ t)$ is completely defined by a relatively small table of coefficients, D_{sk}.

An important advantage of the numerical mapping method is that as new data becomes available the coefficients can be easily revised by the use of an electronic computer. Also, as additional physical relationships are established, they can be incorporated readily into the mapping program. The value of the parameter mapped can be derived for any location, and graphical maps may be

Table 5-1. Geographic Functions, Gk (r, V)

Main Latitudinal Variation		Mixed Latitudinal and Longitudinal Variation			
		First Order In Longitude		Second Order In Longitude	
k	$G_k(\lambda, \theta)$	k	$G_k(\lambda, \theta)$	k	$G_k(\lambda, \theta)$
0	1	$k_0 + 1$	$\cos \lambda \cos \theta$	$k_1 + 1$	$\cos^2 \lambda \cos 2\theta$
1	$\sin \lambda$	$k_0 + 2$	$\cos \lambda \sin \theta$	$k_1 + 2$	$\cos^2 \lambda \sin 2\theta$
2	$\sin^2$	$k_0 + 3$	$\sin \lambda \cos \lambda \cos \theta$	$k_1 + 3$	$\sin \lambda \cos^2\lambda \cos 2\theta$
		$k_0 + 4$	$\sin \lambda \cos \lambda \sin \theta$	$k_1 + 4$	$\sin \lambda \cos^2\lambda \sin 2\theta$
k_0	$\sin^{\theta 0} \lambda$		$\sin^{\theta 1} \lambda \cos \lambda \cos \theta$	K-1	$\sin^{\theta 2} \lambda \cos^2\lambda \cos 2\theta$
			$\sin^{\theta 1} \lambda \cos \lambda \sin \theta$	K	$\sin^{\theta 2}\lambda \cos^2\lambda \sin 2\theta$

prepared by computations using the table of coefficients. The numerical map is particularly useful when large numbers of propagation path computations are required. The use of a computer permits rapid and economical inclusion of all the required propagation variables used in a routine computation. Tables of predicted coefficients may be obtained in the form of tested sets of punched cards from the Prediction Services Section CRPL. Users who do not have access to a computer, or whose requirements are not large enough to justify their own computer facility, may arrange to have ionospheric propagation calculations performed at a reasonable cost by the Frequency Utilization Section of the CRPL. For those who do not have access to a computer, or require a manual solution to an ionspheric propagation problem for some other reason, the CRPL also issues a monthly publication called Ionospheric Predictions. This bulletin contains the basic tables of predicted of coefficients of $f_o F_2$ and $M (3000) F_2$ three months in advance, in addition to world maps of predicted MUF (zero) F_2 and MUF (4000) F_2 for every hour of universal time. The predicted maps are calculated from the tables of predicted coefficients using the approximations:

$$\text{MUF (zero) } F_2 = f_o\ F_2 + f_h/2 = f_x\ F_2,$$

$$\text{MUF(4000) } F_2 = f_0\ F_2 \times M\ (3000)\ F_2 \times 1.1$$

Instructions for the use of the maps may be found in the National Bureau of Standards Handbook 90. Figures 5-11 and 5-12 show examples of world maps of MUF (zero) F and MUF (4000) F_2 for the hour 1800 universal time predicted for June 1963.

Permanent Predictions

In contrast to the monthly prediction systems, permanent predictions systems describe the variation of the ionosphere over a complete solar cycle. They are particularly useful for long term frequency planning and communications equipment specification. Such systems have been issued by the CRPL of the United States, the Radio Research Station of England, the Defense Research Telecommunications Laboratory of Canada, and the Radio Research Laboratories of Japan. These are usually issued in the form of a set of maps for solar cycle minimum and maximum conditions but may also be in the form of a set of maps for a specific level of solar activity accompanied by a set of maps showing the world variation of the dependence on solar activity. Permanent predictions in the form of numerical maps for solar cycle minimum and maximum are available from CRPL and are part of the current CRPL prediction system. By

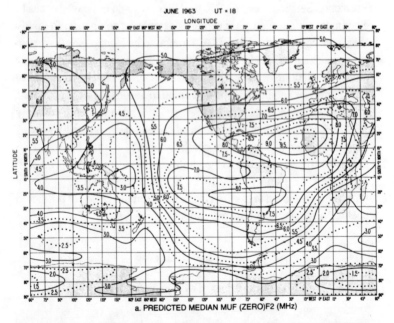

Fig. 5-11. Predicted median F_2 layer parameters for 0800 UCT for June 1963. MUF (zero) F_2. (From CRPL work sheets.)

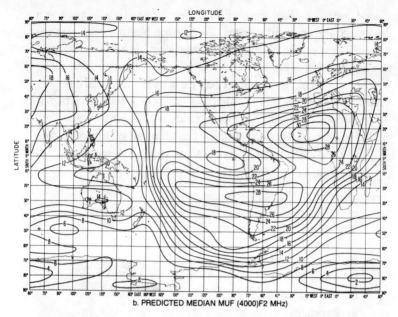

b. PREDICTED MEDIAN MUF (4000)F2 MHz)

Fig. 5-12. Predicted median F2 layer parameters for 1800 UCT for June 1963. MUF (4000) F2.

applying a suitable index of solar activity to a permanent prediction system, ionospheric characteristics may be obtained for any level of solar activity. They may be used as monthly predictions by using a predicted monthly index. The Zurich sunspot number has been the most commonly used index of solar activity. An index based on part ionospheric data, such as I (F 2) of the Radio Research Station, is also sometimes used.

RADIO PROPAGATION FORECASTING

Forecasting of the probable occurrence of an ionospheric storm can be very useful from several points of view. For example, a communication operator is warned to transmit the essential messages before circuits are blacked out; alternatively the operator may start up various relay links, or change the operating frequency. The two radio forecasting centers of the Central Radio Propagation Laboratory regularly issue three types of forecasts of radio propagation conditions. These forecasts apply to sky wave propagation (usually within the frequency range 2 to 30 MHz) over certain important radio paths. The North Atlantic Radio Warning Service is concerned primarily with transmission over paths such as New

129

York-London or Washington-Paris; their forecasts will apply to a lesser extent on nearby paths like Boston-North Africa, Maine-Greenland, etc. The North Pacific Radio Warning Service forecasts are designed for paths like Anchorage-Seattle, San Francisco-Fairbanks, or Tokyo-Anchorage; they can be given special interpretation, however, for others, such as short intra-Alaska circuits, but again with less reliability. Average quality is forecast for a specified 6, 12, or 24 hour period. All times are universal coordinated time (UCT). The CRPL radio quality scale is used in all instances:

1. Useless
2. Very Poor
3. Poor
4. Poor to Fair
5. Fair
6. Fair to Good
7. Good
8. Very Good
9. Excellent

The average quality in the two hours preceding the issue time is expressed in the following letter code:

> W-disturbed (1 to 4)
> U-unsettled (5)
> N-normal (6 to 9)

The forecasts must be interpreted in light of the user's own experience. A forecast of *3* means that the conditions experienced should be relatively poor, but the operator must interpret this in terms of the expected performance of his circuit. If the forecast is correct, no more than 5 or 10 percent of all days should be so poor. There may be, however, a systematic difference between the word descriptions used by the operator and by the forecaster, and the operator must take this into account in interpreting the forecast. For instance, the operator may be used to calling conditions *very poor* when the forecaster calls them *poor* or the operator of another, perhaps easier circuit, may never see conditions as worse than *fair*. The forecasts are expressed on a scale which corresponds to the average experience reported on a typical circuit. The estimate of present conditions (the letter) is made available to aid the user in interpreting the forecast (the number). For example, if an *N-5* is issued, and the operator is already at the time of the forecast experiencing what he would call unsettled (quality 5) conditions, then he would expect conditions to become even worse. The statement *N-5* indicates that the forecaster looks for a deterioration of at least one grade during the forecast period. The value of this scheme rests on the fact that the operator is concerned with expected changes in conditions and so is given information to tie in with his current experience. If the 5 had been issued without the *N*, the user would not have been prepared for a drop in quality since he was already experiencing what he called quality 5 conditions, although the consensus available to the forecaster rated conditions *normal*.

A third forecast, *the 24 hour forecast*, is also available daily by telephone and teletype from two forecasting centers. Separate quality estimates are issued for nighttime and daytime of the ensuing 24 hour period. The range of frequencies on which predicted quality should be realized is also included. Additional information concerning the radio forecasting services may be obtained from:

For the North Atlantic Area:
North Atlantic Radio Warning Service (or NARWS)
National Bureau of Standards
Box 178
Fort Belvoir, Virginia 22060
(Telephone-Washington, D.C. 780-1436 or 780-1444)

For the North Pacific and Alaskan Area:
North Pacific Radio Warning Service (or NPRWS)
National Bureau of Standards
Box 1119
Anchorage, Alaska 99510
(Telephone-753-2211 or 753-7210)

Past CRPL forecasts together with actual conditions which developed are included in the CRPL Fseries, B, Solar Geophysical Data.

Chapter 6
Station Accessories

Eventually the serious amateur/experimenter may want to investigate further into the mysterious and fascinating world below the broadcast band. The aim of this chapter is to acquaint the reader with readily available LF and VLF commercial receivers, ranging from ultra sophisticated research units to military receivers well within the budget of the average experimenter. Also included are plans for two LF and VLF converters that can be used with either a standard short wave receiver or with an amateur communications receiver. Included too is a time constant integrator that can be used to graphically record the signals.

COMMERCIALLY AVAILABLE RECEIVERS

An experimenter able to obtain a commercial receiver is lucky indeed, for these packages represent the ultimate in tuning accuracy, sensitivity, and other important characteristics.

Tracor VLF/LF Model 599K Tracking Receiver

The Tracor model 599K VLF/LF tracking receiver is the ultimate in a superbly designed piece of electronic gear. It is being used by many colleges for research studies in propagation, including the diurnal change in altitude of the VLF reflecting layers, studies of the effects of solar disturbances, and comparison of reception of various VLF transmissions. The 599K receiver has also been very useful in timing applications that involve maintaining long term synchronization of "clocks" at separate locations. Each receiving station can operate from the same frequency and time base by monitoring a selected VLF transmitter. Thus, initial synchronization of two or more timing systems can be maintained indefinitely. The receiver may be also used to calibrate any frequency standard with a 100 kHz or 1 MHz output. The relative frequency error between the local standard and the received VLF carrier signal will be observable as a phase drift on the microseconds digital counter and an associated chart recorder, as seen in Fig. 6-1. The rate of this phase drift can be interpreted directly as a fractional frequency error in the local standard; thus a phase rate of one microsecond per 100 second time interval represents a fractional frequency deviation of one part in 10^8. Single sideband techniques have been used in the receiver rf section to eliminate the need for rf filters. The unique image canceling properties of this technique reject images at least 60 dB, below 30 kHz, 50 dB above, with no rf filtering of any kind. The receiver is continuously tunable in 50 Hz increments from 3.00 kHz to 99.50 kHz. A rundown of important specifications follows.

Frequency Coverage. Stabilizes carrier tracking over the range from 5.0 kHz to 99.5 kHz. This is accomplished by the use of a thumb wheel switch which selects tens, units, and tenths of kHz, and 0 or 50 Hz. The receiver itself is usable down to 3 kHz.

Fig. 6-1. The Tracor Model 599K VLF/LF Tracking Receiver. Photo courtesy Tracor Electronics.

Receiver Sensitivity and Signal/Noise Performance. A noise figure of 0.01 microvolt signal (corresponding to 2.0 microvolt/meter field strength at 20.0 kHz with the Model 599-603D loop antenna) from 5 to 30 kHz and 0.02 microvolt signal from 30 to 99.95 kHz into the receiver energizes the carrier level switch and enables normal phase tracking; this tracking is maintained at an input signal-to-noise ratio of 50 dB. (Gaussian noise is measured in a 1 kHz bandwidth.)

Calibration Accuracy. Short and long term stability is better than ±0.25 microseconds relative to received carrier, under typical laboratory conditions. Calibration accuracy is better than $±1 \times 10^{-11}$, when averaging for 24 hours.

Frequency Standard Input. The receiver accepts any stable 1 MHz or 100 kHz signal from an external frequency standard.

Recording Outputs. Outputs are available for recording phase and coherent carrier amplitude information. (Phase output is a voltage proportional to relative phase difference between the local standard and the VLF carrier.) Two of these outputs are available simultaneously to allow recording on a 1 mA recorder with 10 and 100 microseconds full scale deflection. The coherent carrier amplitude output is derived from the receiver's agc voltage and has an approximately logarithmic characteristic over a 40 dB range. Carrier information may be recorded on any 1 mA recorder.

Recorder. Terminals are provided for an external chart recorder (the recorder on the front of the receiver is usually an option). The small recorder, if supplied, has a full scale of either 10 microseconds or 100 microseconds.

LO Synthesizer. Any appropriate LO frequency, phase normalized, is generated by a coherent synthesizer system. The main tuning is accomplished by a series of thumbwheel switches that control and indicate frequency of reception. A choice of local oscillator 1 kHz above or below signal frequency can be accomplished by switching located on the front panel. The phase of the local oscillator signal is absolutely fixed by the synthesizer control circuits. Average LO phase recovers within 0.05 microseconds of its original value, when synthesizer tuning is changed and then restored to its original setting.

AGC, and Dynamic Range. The stable agc circuit used in the receiver assures full-reliability phase-locked servo operation over a

40 dB range of carrier level. Total variation of phase due to agc alone is less than 0.3 microseconds. The total signal level operating range is 120 dB, including 40 dB agc, and 80 dB mgc. The range of acceptable signals is 0.01 microvolt to 10 mV at the antenna input terminal. Interfering signals removed more than 200 Hz from the tuned frequency and no larger than 100 millivolts cannot cause rf or i-f saturation.

RF Selectivity. The circuitry of the receiver is so designed that it eliminates all need for narrow band rf preselection or band switching of filter networks, without degrading performance.

The image rejection is at least 60 dB from 10 to 30 kHz and at least 50 dB above 30 kHz.

Bandwidth. The range of a broad band, low gain receiver front end amplifier extends from below 3 kHz to approximately 100 kHz.

Antenna Requirements. The receiver can be used with *any* loop whip or simple long wire antenna. Nominal antenna input impedance is 50 ohms.

Power Requirements. The receiver operates from 95–125 Vac, 48–62 Hz, or from +12 and –12 Vdc on external standby batteries.

The Tracor Model 599K VLF Tracking Receiver is an exceptionally fine piece of electronic gear for research use, and engineered in such a way as to avoid obsolescence. The price is in the $4,000 range; however, used receivers can be bought in rebuilt condition for about $1,000. Tracor also manufactures a Model 900 VLF/LF receiver complete with antenna (9.9 kHz to 25.6 kHz, and 59.9 kHz to 75.6 kHz) and recorder for $1,000. These receivers can be purchased in rebuilt condition in the $500–$800 price range. Although the Model 900 lacks some of the refinement and accuracy of the 599K it offers the *serious* experimenter a good research instrument with a moderate investment.

The National Radio HF/VLF HRO-500 Receiver

The National HRO-500, shown in Fig. 6-2, is a completely solid state communications receiver with a design that includes the complete VLF/HF spectrum between 5 kHz and 30 MHz. The receiver uses all facilities for a-m, cw, and ssb reception, with an unusually accurate dial calibration, high frequency stability, and ease of operation. The HRO-500 is currently in use by major government agen-

Fig. 6-2. The National Radio HRO-500 HF/VLF communications receiver. Photo courtesy National Radio Company.

cies, colleges, and serious experimenters for HF/VLF point-to-point communications, monitoring, and laboratory instrumentation applications. Completely transistorized the HRO-500 generates minimal heat gradients, and reduced power requirements allow the receiver to be used in portable or field applications barred to existing vacuum tube equipment. The HRO-500 is remarkably compact in size and may either be table or rack mounted. In addition a fully portable version is available with a self contained power supply, loudspeaker, and carrying case. Important features and specifications follow:

1. The HRO-500 covers the entire VLF/HF spectrum between 5 kHz and 30 MHz.
2. The receiver uses total transistorization for maximum stability.
3. Complete ssb, a-m, mcw, and cw facilities, including both product and diode detectors. Pass band tuning of the filter for true selectable side band reception: fast attack, slow decay, agc and four discrete bandwidths from 800 Hz to 8 kHz.
4. Dial calibration is accurate to one kilohertz over the entire frequency range, with bandspread of 1/4 inch per kilohertz and 24 feet per megahertz.

5. Selectable tuning ratio of either 10 kHz or 50 kHz per revolution of main tuning control for choice of either rapid band scanning or fine tuning at will.

6. Superb short and long term stability, Nominal stability including ±27% change in ac input voltage, better than 100 Hz per day and 50 Hz per °C.

7. Frequency determination by means of phase-locked crystal frequency synthesizer gives maximum stability and eliminates band-to-band recalibration.

8. Sensitivity and image rejection (averages 80 dB, minimum 50 dB).

9. Flat audio response—100 Hz to 5500 Hz within 6 dB, over with two watts of available audio output.

10. Frequency stability in room ambient temperature is 300 Hz per hour from point of turn-on to two hours after turn-on; better than 100 Hz day thereafter.

11. Selectivity is accomplished by the use of 6 dB bandwidths available: 500 Hz; 2.5 kHz; 5.0 kHz; 8.0 kHz. Filter design is of the six pole LC type operating at 230 kHz: tunable through 6 kHz range in 500 Hz and 2.5 kHz bandwidths.

12. Antenna inputs are of 50 ohm unbalanced for HF, and high impedance unbalanced for LF. The HRO-500 has been used by amateurs and SWLs for a great number of years. Earlier tube versions of this receiver can be purchased in the $500-$1,000 price range depending on age and condition. Although the HRO-500 has no provisions for recorder outputs, the 3.2 ohms available at the speaker terminals is more than sufficient to drive the integrator described in this chapter.

THE DRAKE MODEL DSR-2 RECEIVER

The Drake Model DSR-2 is another high grade communications receiver employing most of today's up to date solid state devices and circuitry. It provides continuous coverage from 10 kHz to 30 MHz. The received frequency is displayed on six Nixie tubes (Fig. 6-3) to the nearest 100 Hz. Frequency injections of the DSR-2 are controlled by a phase-locked digital synthesizer which allows incremental

Fig. 6-3. The Drake DSR-2 communications receiver. Photo courtesy RL Drake Company.

frequency selection of 10, 1, and 0.1 MHz steps. The remaining 0 to 0.1 MHz is continuously adjustable by the use of a variable oscillator controlled from the fine tuning knob on the front panel. Modular construction on readily accessible printed circuit boards is used throughout the DSR-2. The extensive use of dual gate MOS-FET transistors in the DSR-2 circuitry contributes to good intermodulation, avc, wide dynamic range and overload performance. The front panel is designed so that the operator can easily select frequency (with the fine tune control) a-m or ssb product detector, i-f bandwidth, AF gain, and bfo in seconds. The independent sideband (ISB) is a built-in feature. Seperate i-f crystal filter, i-f amplifier, and audio output circuits allow two simultaneous communicaton channels to be employed on one frequency assignment, doubling the information receiving capacity. Front end protection includes special circuitry built in to provide protection against transmitters in close proximity. It will withstand a 30 volt emf from a 50 ohm source, with the receiver on or off. The normal avc system has appropriate attach and decay times to provide proper clean reception of ssb and cw signals. Important specifications are:

Frequency Range—10 kHz to 30 MHz continuous.
Modes of Operation—usb, lsb, cw, RTTY, a-m, isb.
Frequency Readout—Complete to 100 Hz on six Nixie tubes.
Frequency Selection—10 MHz, 1 MHz, 0.1 MHz steps switch selected. 0 to 0.1 MHz continuous.

Frequency Stability—Frequency drift does not exceed 200 Hz in any 8 hour period at a constant ambient temperature between 0° and 40° C and ±10% variation from nominal line voltage after 1/2 hour warmup.

Sensitivity—0.01–0.5 MHz: Less than 4 microvolts for 10 dB SINAD at 2.4 kHz ssb mode. Less than 25 microvolts for 10 dB SINAD at 6 kHz a-m mode with 30% modulation. 0.5-30 MHz: Less than 0.3 microvolts for 10 dB SINAD at 2.4 kHz ssb mode. Less than 2 microvolts for 10 dB SINAD at 6 kHz a-m mode with 30 % modulation.

Antenna Input Impedance—10 kHz to 500 kHz 1000 ohms
500 kHz to 30 MHz 50 ohms

Although the R.L. Drake DSR-2 is a relatively new LF/HF receiver, a few used ones are appearing on reconditioned equipment lists. Prices range from $500 to $750, depending upon the age and condition of the receiver.

MILITARY (SURPLUS) LF AND VLF RECEIVERS

Military equipment offers excellent values in receivers for use in solar flare recording. These items are usually of excellent quality too.

The Model RBA-7 Receiver

The Model RBA-7 radio receiver is a long range, long wave communications receiver designed for use aboard ship or in land based stations. The receiver covers the range of 15 to 600 kHz and is adapted to receive A1, A2, and A3 emissions (cw, mcw, and voice). Selectivity of the receiver is such that voice reception will be of low quality and not usable below 300 kHz. These receivers are ideal for propagation and flare recordings due to a number of reasons. First, they are readily available from surplus electronic dealers at reasonable prices. (The receiver as shown in Fig. 6-4 was purchased by the author for $150, in mint condition.) Second, they are selective enough to filter out anything but the incoming signal, and third, they have a series of 600 ohm outputs that can be used to drive any number of recorders, using time constant integrators. The strength of the audio signal is also enough (60 microwatts) to sufficiently drive a 0 – 100 mA recorder to full scale, using no rectification whatsoever. With all of these advantages in mind, it would be safe to say that this is the ideal LF/VLF receiver for the amateur/

Fig. 6-4. The author's RBA-7 (15 to 600 kHz) military communications receiver.

experimenter; therefore, we will devote a full description dealing with the operation of this excellent receiver.

General Description. The major units of the RBA-7 equipment are the Navy type CFT-4615B receiver and a Navy type CBOG-21030-D rectifier power unit. The two units are housed in separate shelf-mounting cabinets and are connected together by means of appropriate cable and plugs. The receiver will operate from either a single wire antenna feed system or a coaxial line. The design is such that several receivers may be operated from one common feed system. The receiver is commonly used with headphones, any number from one to twenty may be connected to the receiver output. The power unit for the receiver operates from 110, 115, or 120 volt, 55 to 65 Hz source to deliver all of the required voltages. The power unit is so designed that two receivers may be operated from a single power unit. Operating potentials under these conditions are slightly lower.

The RBA radio receiver equipment has undergone minor component changes since its initial introduction as the Model RBA. Models RBA through RBA-3 are identical except for nameplate changes due to different contractual data. The contract for the

Model RBA-4 receiver was canceled. The Model RBA-5 is identical to its predecessors except that a few component values were changed to conform with standard JAN nominal sizes. The Model RBA-6 is identical to the RBA-5 except that it is provided with a rack mounting panel instead of a cabinet. The Model RBA-7 is essentially the same as the Model RBA-5 except that wherever practical resistor and capacitor sizes are changed to standard JAN values, and other minor manufacturing changes are made. With these facts in mind, the RBA-7 instruction and maintenance manual may be applied to all RBA receiver equipment. Components of the various models are interchangeable. The use of W-type tubes as specified for the RBA-7 is preferable for all models.

Theory of Operation. The Model RBA-7, Navy type CFT, 46154B, radio receiver is an eight 'tube, tuned radio frequency communications receiver. It covers all radio frequencies from 15 to 600 kHz. This tuning range is divided into four switch selected bands with a five percent overlap between adjacent bands. The receiver is functionally divided into a three stage tuned rf amplifier, a one tube heterodyne oscillator, a triode detector, and a three stage audio af amplifier. The heterodyne oscillator is employed to produce an audio beat tone for the reception of cw signals. The functional block diagram Fig. 6-5 shows the arrangement of the receiver. The explanations which follow make reference to individual simplified schematics.

Tuned RF Amplifier. The three-stage tuned rf amplifier utilizes four sets of tuned-band switched coils; each set covers one of the four frequency bands. The coils used in each set are the antenna, first rf, second rf, and third rf stage coils. Therefore there are 16 separate coils employed in the rf tuning portion of the receiver. The receiver input is designed for connection to either a single wire antenna system or to a coaxial feed line from a loop type antenna. By referring to the simplified schematic diagram in Fig. 6-6, you will note that the circuit is adjustable for peak performance with a wide variety of antennas. This same adjustment feature makes connection of several receivers to a common antenna system feasible. The signal is fed into the receiver via coaxial input jack J-102 and coupled to a tap on the input coil. The antenna coil Q for the two lowest band coils is extremely high, as it also is on the low band coils of the other rf stages. High Q circuits characteristically possess a great deal of

141

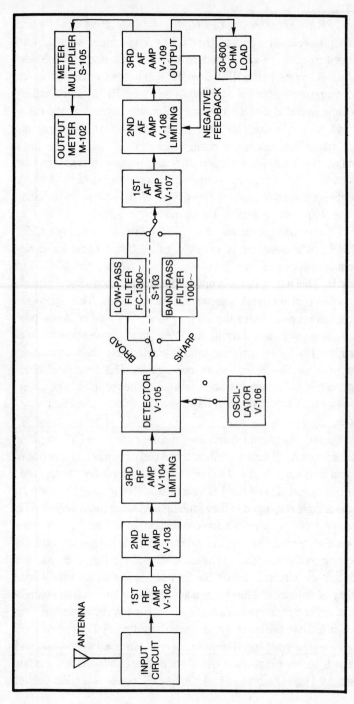

Fig. 6-5. Functional block diagram of the RBA-7 (LF/VLF) receiver.

142

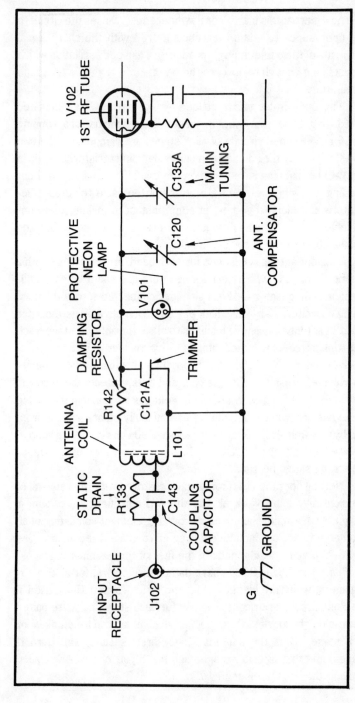

Fig. 6-6. Simplified schematic diagram of the RBA-7 input circuit.

143

electrical inertia; this inertia, or flywheel effect, causes the circuit to ring upon shock excitation (as encountered with the sharp signal transients of radio telegraphy), producing a damped oscillation which persists long enough to obscure the on-off state of a radio telegraph transmitter.

This undesirable effect, caused by high circuit Q, is eliminated by reducing the Q to an optimum value, through the use of a damping resistor. The tuned circuits of all rf stages are damped, on the two lowest bands, by the addition of a series damping resistor, as seen in Fig. 6-6; no damping is required on the two highest bands. Without damping the signal elements of transmitted radio-telegraph elements would blend together, in some instances making them unreadable. Static "crashes" would persist as a ringing sound and further obscure the received signal.

The four antenna coils, one for each band, are each fitted with a trimmer capacitor which permits separate alignment of each coil. Antenna compensator C-120 is a padder connected across the main tuning capacitor, C-135A, which permits corrective adjustment for different antenna capacitive loadings to make the antenna stage track with the succeeding tuned stages. This padder is a front panel control labeled ANT. COMP. A protective neon lamp, V-101, limits the voltage across the antenna coil, and thus protects the receiver against damage from excessive input voltages such as might result in the event of strong static bursts or unusually strong signals from nearby transmitters. Voltages in excess of 90 volts cause the lamp to flash, effectively short circuiting the input circuit for only as long as the strong signal persists.

Normal operation is automatically restored with the removal of the excessive input voltage. The three rf amplifiers are of similar design; each employs an identical coupling element consisting of an untuned primary coil inductively coupled to a tuned secondary. See Fig. 6-7, which is a schematic of the first of three similar rf stages. The band 1 and band 2 coils have damping resistors, for all three rf amplifiers, which are identical in function and design to those used in the input circuit. Four coils are used for each stage; coils are paired in common shield cans, there being two shielded can assemblies for each stage. Thus the antenna stage, first, second, and third rf stages, and the oscillator stage each has a pair of coil assemblies, bringing the total number to 10.

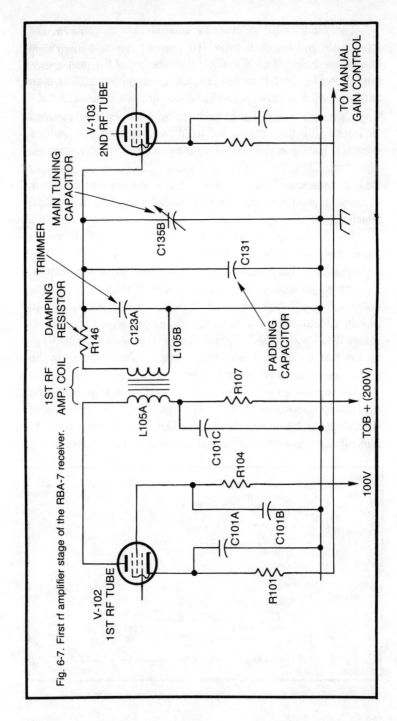

Fig. 6-7. First rf amplifier stage of the RBA-7 receiver.

145

The band switch selects, by separate set of contacts, the appropriate coil for each band. To prevent resonant absorption effects, the band switch also shorts out the coil of the next lowest band. Refer to Fig. 6-8 for band switch operation details. The main tuning capacitor is a five section ganged control which tunes all of the stages simultaneously. The individual stages are separately aligned for tracking by their respective fixed padder capacitors, and the individual coils are likewise aligned by their adjustable trimmer capacitors. The tuning dial has separate scales for each band but only the applicable scale is in view due to the action of a traveling mask which is moved by means of a mechanical connection to the band switch.

The dc voltages applied to each stage are supplied through decoupling resistors R-107, R-108, and R-109 (refer to Fig. 6-9 (a) and 6-9 (b)), and rf filtered through a series of bypass capacitors. The dc voltages applied to the first and second rf stages are identical. Cathode bias is developed across resistors R-101 and R-102. An additional, controlled bias, voltage is developed across variable resistors R-128 and R-136A. These volume controls operate by varying the bias on the first and second stages, thereby varying the mutual conductance of the two tubes and controlling their gains.

Resistor R-128 is a special volume control that is geared to the main tuning capacitor so it decreases the gain as the receiver is tuned towards the high frequency end of the dial. This decrease tends to automatically offset the increase of receiver gain associated with a

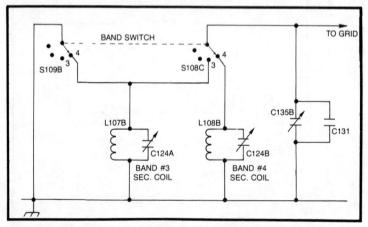

Fig. 6-8. Band switch operation of the RBA-7 receiver for band 4.

reduction of tuning capacitance. The overall gain of the receiver is thus held to a fairly constant level. The main rf gain control, R-136A, is coupled to audio gain control R-136B so that audio and rf gains are adjusted by the same control and maximum signal-to-noise ratio is obtained at all times.

The gain control action of the two potentiometer sections is essentially consecutive rather than simultaneous. The two tapers have been deliberately selected to make the first 70 percent of clockwise rotation (0 to 70 on the dial scale) act principally to increase rf gain while the remaining 30 percent (70 to 100 on the dial scale) act primarily to increase af gain. In this manner tube microphonics receive maximum amplification only when the gain control is fully advanced. Thus noises in the receiver are amplified only to the same relative degree as the signal, keeping the signal-to-noise ratio always optimum.

The dc voltages applied to the third rf amplifier are considerably lower than those applied to the first two stages. Plate and screen potentials of approximately 15 volts limit the maximum output voltage of this stage to prevent overloading the detector. With voltages of this low order it is not practical to vary stage gain control of the bias voltage; therefore, the cathode of this stage is returned to ground through its bias resistor, R-103, without being connected to the volume control circuit.

The behavior of this third rf stage is *not to be confused with automatic volume control*. This circuit is purely an output limiter which behaves as it does because of its inherent inability to deliver more voltage to its output circuit than the dc that is available to its plate. This is one of the features that make the RBA-7 ideal for SES and SEA studies. Since no avc is employed in the circuit, the output to the recorder is an exact trace of the strength of the incoming signal.

The heterodyne oscillator operates at a frequency of 1,000 Hz above the resonant frequency of the other tuned circuits, producing a signal which is mixed with the received cw carrier as it comes from the output of the rf amplifier. The resultant 1,000 Hz difference product is later demodulated, in the detector, as an audible note. The frequency of this electron-coupled oscillator is governed by one of four coils, which is switched into the circuit by the band switch, and by a section of the main tuning capacitor. The tuned frequency of

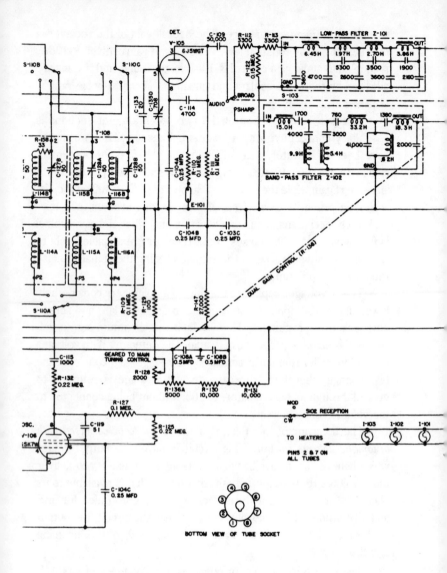

Fig. 6-9(a). The RBA-7 schematic for antenna input circuits through the 3rd i-f stage.

the rf amplifier is thus automatically tracked to produce the same beat tone (1,000 Hz ±0.75% of the signal frequency) for all received signals. Resonant absorption effects are prevented in this stage, as in all other stages, by shorting out the unused coil of the next lower band. The output of the heterodyne oscillator is partially isolated, by

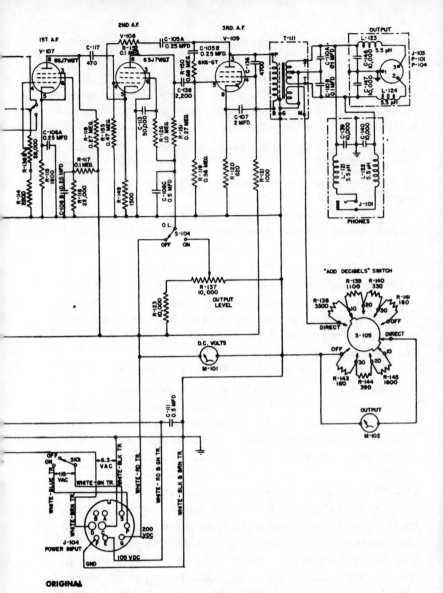

Fig. 6-9(b). The RBA-7 schematic for the detector stage through the audio output stage.

means of resistor R-132, to prevent undue loading, and fed to the plate circuit of the third rf amplifier switch S-102, permitting the oscillator to be disabled for the reception of modulated signals.

Circuit features which depart slightly from the basic electron-coupled oscillator will be revealed by a study of the schematic

149

diagram. It will be noted that the tuned elements of the various bands are not schematically identical. Each band has been individually adjusted by means of series or parallel capacitors for the correct LC ratio required for close tracking with the received signal. Another feature of this oscillator is output equalizing capacitor C-119, which partially compensates for the increased output normally occurring as the oscillator is tuned to the high end of each band. As the frequency increases, the impedance of C-119 decreases, therefore lowering the oscillator plate load impedance and decreasing the oscillator output.

You may also note that the plate and screen voltages for the oscillator are supplied from the regulated 105 volt bus. The output of the third rf amplifier then mixes with the heterodyne oscillator output, when required, and passes directly into the detector. A triode detector, biased approximately to cutoff by means of a large value of cathode resistance, accomplishes the required demodulation by plate rectification. Plate current of this stage is varied only by the positive rf excursions. Thus rectification and amplification both take place at the same time.

The detector output is then modified by one of two wave filters, Z-101 or Z-102, to remove signals which do not contribute to the intelligence of the received communication. These filters contain both broad and sharp audio response characteristics. The sharp characteristic is achieved by using a very narrow bandpass filter with a center frequency of 1,000 Hz. This filter is particularly useful for reception of cw signals as it will pass the 1,000 Hz beat tone to the practical exclusion of all other signals and noise.

The broad characteristic obtained from the output of a sharp cutoff low pass filter is useful for reception of voice, as well as radiotelephone transmissions. The audio amplifier of the RBA-7 receiver amplifies the detector output for delivery to the headsets. It is a three stage voltage and power amplifier with two special features. Output level control R-137 permits the overall audio output to be held at any desired level despite changes of signal strength or modulation percentages. The behavior of this circuit should not be confused functionally with that of the third rf stage which limits the maximum carrier output presented to the detector. The output limiter control is to be adjusted by the operator to prevent high signal levels, which may occur at the louder peaks encountered when receiving fading signals, from blasting in the headphones. The fixed limiting inherent in the third rf stage is provided to prevent overloading the detector.

The second special design feature in the audio amplifier is the constant-voltage output system which permits any number of headphones, from 1 to 20, to be switched in and out at will and without noticeably changing the volume level in those headphones still in use. This is accomplished by the use of inverse feedback using three audio amplifier stages. The output circuit of the audio amplifier is provided with two output windings, one for matching the power output to the headset and the other for providing an adequate voltage to operate the volume indicator. Both secondary windings are electrostatically shielded from the primary to attenuate any rf which might be picked up in the receiver from the long output lines.

Further protection from rf interference is provided by filters in the output connections of the load winding. The output load winding is brought out, through separate rf filters, to a headset jack and a three contact socket for connection to the output line. This winding is center tapped and balanced to ground. Output wiring and headset connections must therefore *not* be grounded. The meter output winding does not have as great a step-down ratio as the load winding and thus the voltage available to the meter is made sufficiently high to provide adequate meter deflection, even with low audio output. The meter may be made less sensitive when using higher output levels, by means of an ADD DECIBELS switch located on the front panel. Meter multiplication factors corresponding to 0, 10, 20, or 30 dB may be selected by this switch. The 0 dB position, labeled DIRECT, is spring loaded so that the switch must be hand held in this position to obtain maximum meter sensitivity.

The RBA-7 receiver can be found in use in any number of places. These range from industrial research laboratories and physics laboratories at major colleges to instrumentation calibration and repair shops. Although the receiver is relatively old (most units were reconditioned around 1959), it is stable and reliable, and well worth the $150 to $200 price. Refer to Appendix B for the surplus dealers' addresses.

The Rycom Model 2174 Voltmeter Receiver

The Rycom 2174 is a commercial surplus receiver covering 0 to 420 kHz in six bands. It was originally designed for low frequency work by the government. The bandwidth switch of this receiver selects 100 Hz, 3 kHz, and 10 kHz. It is selectable for AM, LSB, USB, BF01, and BF02 signals. Antenna inputs are for balanced or unbalanced lines of 135 ohm, 600 ohm. The receiver includes a built in speaker and monitor jack, and can be either rack or table mounted. Price range: used-repairable, $95; or in a repaired condition for about $125. Refer to Appendix B for the addresses of the surplus dealers.

Although there are numerous other LF-VLF receivers on the surplus market, most of these are outdated units manufactured during World Wars I and II. A few of these receivers show up in the surplus circulars, but they lack some of the sensitivity and selectivity of the aforementioned receivers. Also they suffer from one common affliction which the author calls "hardening of the arteries." It simply means that the wiring is aged and can easily be broken. Such receivers have to be rewired and, along with the problem of obsolete tubes, makes them not worth the time and effort required to put them in operating condition.

LF AND VLF CONVERTERS

Many amateurs/experimenters own either general coverage receivers or ham band only receivers. These receivers can be modified to cover LF and VLF bands by the use of converters. The two converters about to be described are easily built and require a modest investment.

A Tuned VLF Converter for General Coverage Receivers

This novel tunable inductance approach provides the basis for a tuned VLF converter that covers a wide frequency range without bandswitching. One of the big problems the amateur/experimenter may find in trying to build a converter for this part of the spectrum is in constructing a variable tuned circuit which will cover a substantial portion of the desired frequency range. Assuming that the desired band extends from 10 to 150 kHz, with a ratio of the corner frequencies of $150:10 = 15$, the tunable component must have a variation of $15^2 = 225$. Since this cannot be successfully accomplished with conventional variable capacitors or even inductors, the frequency range has to be divided into a number of subbands or the tuned circuit is eliminated altogether. The latter is done in most converters; they are untuned.

The Tuned Circuit

There is, however, a novel method of inductive tuning which will cover the required range. This method makes use of a toroidal ferrite core which is magnetically biased by a pair of small permanent magnets as shown in Fig. 6-10. By rotating one of the magnets with respect to the other, the amount of flux penetrating the toroid is varied, changing the ferrite's permeability and in turn the inductance. It is interesting to note that maximum flux penetration and minimum inductance occur when like poles are opposite one another. The two magnets used to bias the toroid inductor are of the *button* type with a one half inch (13 mm) outside diameter. (Available from Edmund Scientific Company, refer to Appendix B.) The outside diameter of the toroid is also one half inch. The whole tuning assembly is built around the shaft of a discarded potentiometer, as shown in Fig. 6-11. The particular toroid the author used required 100 turns of stranded wire for an inductance variation of 100 μH to 12 mH (a 120:1 range). However, ferrite cores with higher permeability would require fewer turns. Measured Q values for this inductor were around 50 (using a Boontoon Q meter) for frequencies between 10 and 150 kHz.

Converter Circuit

The circuit of the VLF converter, which has an output on 15 meters for use with a communications receiver, is fairly conventional as shown in Fig. 6-12. The antenna is coupled directly to the "hot" end of the tuned circuit (or through a capacitor to provide a degree of

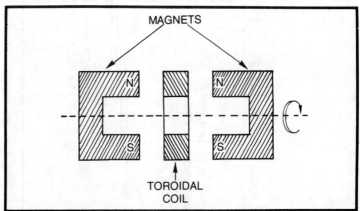

Fig. 6-10. A method for magnetically biasing a toroidal ferrite core with two small butt on magnets.

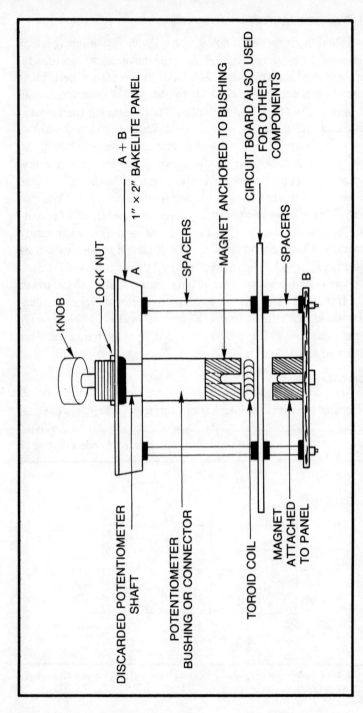

KNOB

LOCK NUT

A + B
1" × 2" BAKELITE PANEL

A

SPACERS

MAGNET ANCHORED TO BUSHING

CIRCUIT BOARD ALSO USED
FOR OTHER
COMPONENTS

SPACERS

B

DISCARDED POTENTIOMETER
SHAFT

POTENTIOMETER
BUSHING OR CONNECTOR

TOROID COIL

MAGNET
ATTACHED
TO PANEL

Fig. 6-11. The method used for constructing the magnetically biased tuning inductor used in the VLF converter.

154

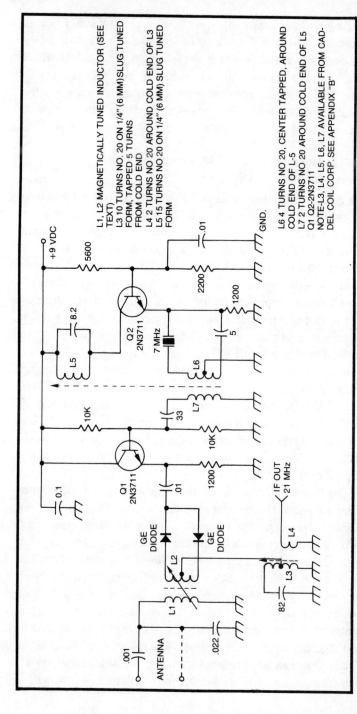

Fig. 6-12. Schematic diagram of the tuned VLF converter. The local oscillator uses a 7 MHz FT-243 crystal in the third overtone mode.

L1, L2 MAGNETICALLY TUNED INDUCTOR (SEE TEXT)
L3 10 TURNS NO. 20 ON 1/4" (6 MM) SLUG TUNED FORM, TAPPED 5 TURNS FROM COLD END
L4 2 TURNS NO 20 AROUND COLD END OF L3
L5 15 TURNS NO 20 ON 1/4" (6 MM) SLUG TUNED FORM
L6 4 TURNS NO 20, CENTER TAPPED, AROUND COLD END OF L-5
L7 2 TURNS NO 20 AROUND COLD END OF L5
Q1 Q2-2N3711
NOTE-L3, L4, L5, L6, L7 AVAILABLE FROM CAD-DEL COIL CORP. SEE APPENDIX "B"

155

matching to long antennas). The mixer uses a *matched* pair of germanium diodes, and the local oscillator uses a FT-243 variety third overtone crystal. To obtain third overtone oscillation at 21.000 MHz, choose a crystal with a fundamental frequency a few kHz above 7 MHz. By simply changing the crystal frequency and the oscillator and i-f output coils, output can be changed to any desired frequency band.

A Simple Low Frequency and Broadcast Converter

This converter is designed for use with a ham band communications receiver. Another bonus is the ability of the converter to tune and give continuous coverage from 1600 to 10 kHz, including the broadcast band. This rather unusual range is accomplished without the need for large coils usually associated with a conventional VLF receiver. The circuit, shown in Fig. 6-12, is a standard converter circuit consisting simply of an oscillator and a mixer. Several different transistor types can be used successfully. The only requirement is that the transistor be suitable for rf or oscillator use in regular broadcast receivers. The i-f can be anything between 1.8 and 2 MHz. The author used 2 MHz because it was readily available, and it is the quietest frequency in the range at his location. The oscillator portion of the converter operates over the small range of 2010 to about 3600 kHz. But with a 2 MHz i-f, this gives a signal range of 10 to 1600 kHz. In the conventional receiver circuit usually used for low frequency reception, tuning this range without changing coils would require a tuning capacitor having a maximum-minimum capacitance range of over 25,000 to 1.

To be sure, some rather large inductance and capacitance values are needed in the mixer input circuit, but the values are not critical. A fixed capacitance may be used for the VLF range, and rf chokes of various types can supply the needed inductance in compact form. The complete unit, batteries included, was constructed on a $4 \times 4 \times 2$ inch utility box fitted with a small aluminum panel extending above it. A broadcast band tuning capacitor was used at C5 because of its low cost, although it provides a tuning range much greater than is needed or desired. This range was reduced by inserting fixed capacitor C4 in series. The coil used for the oscillator L5 is a broadcast band antenna loopstick. Its wide-range slug adjustment assists greatly in trimming the circuit to cover the desired frequencies. Keeping the rf input circuit peaked up over such a wide frequency range can also present a problem. However, by switching smaller inductances in parallel with L3 for the higher frequencies, the input circuit can be adjusted sufficiently close to the operating frequency to provide satisfactory reception.

Construction

Coils L3 and L4 originally comprised a two-pi rf choke found in a surplus tuning unit. The pi used for L3 has less inductance than the other, but it is larger physically because it is wound with larger wire. This coil measures 1-1/2 inches in diameter and 1/2 inch thick. The two coils are spaced 1/8 inch apart on a ceramic pillar. The ground connection is made to the lead between the two pi's. A suitable substitute is a conventional three-pi choke of about 10 mH. (Miller 4672). With this choke, the ground connection should be made to the lead between the center pi and either end pi. The single pi is then used for L3, and the remaining two pi's in series constitute L4. Other suitable coils are Miller type 640 rf choke (2.5 mH) for L3, and type 660 (7.5 mH) for L4. These should be mounted, back to back, as close together as possible. L1 and L2 may be almost anything you can find that can be adjusted to the approximate inductance specified by removing turns or adjusting a slug. For the amateur/experimenter who is not familiar with coil winding and inductances the American Radio Relay League L/C/F Calculator is available. It simplifies the matter of getting the correct coil length, number of turns per inch, and coil diameter. Consult the Appendix for other details on the calculator.

In the author's prototype, with the antenna connected directly to L3, the range of 200 to 500 kHz suffered from broadcast station hash. Switching in series capacitor C1 provided much quieter operation and sharper peaking over this range. Above 500 kHz, there is not much difference in reception with C1 in or out of the circuit. Below 200 kHz, signals are stronger with C1 shorted out. The 0.01 μF capacitor, C3, resonates the antenna circuit in the vicinity of 30 kHz, which proves to be satisfactory over the VLF range. Peaking capacitor C2 has little effect on the tuning over this range, of course, because of its small value as compared to the value of C3. The fixed capacitor also works as a good bypass for higher frequencies, eliminating a number of spurious signals from strong stations in the

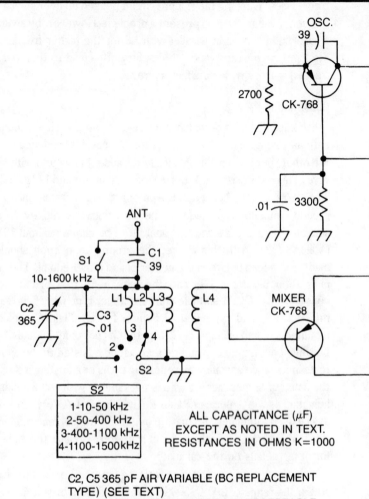

OSC.
39
2700 CK-768

.01 3300

ANT
S1
C1 39
10-1600 kHz

C2 365
C3 .01
L1 L2 L3
3
2
1
4
S2
L4
MIXER CK-768

S2
1-10-50 kHz
2-50-400 kHz
3-400-1100 kHz
4-1100-1500 kHz

ALL CAPACITANCE (μF)
EXCEPT AS NOTED IN TEXT.
RESISTANCES IN OHMS K=1000

C2, C5 365 pF AIR VARIABLE (BC REPLACEMENT
TYPE) (SEE TEXT)
J1 PHONO CONNECTOR OR COAX RECEPTACLE
L1 APPROX 730 μH (MILLER TYPE 620 RF CHOKE
OR TYPE 4412 SLUG TUNED COIL)
L2 APPROX 70 μH (MILLER TYPE 72F68 5AP RF
CHOKE OR TYPE 4409 SLUG TUNED COIL)
L3 APPROX 2.2 mH (MILLER TYPE 640 RF
CHOKE, OR SEE TEXT
L4 APPROX 6.8 mH (MILLER 660 RF CHOKE, OR
SEE TEXT
L5 BC BAND LOOP ANTENNA SLUG-TUNED
(BURSTEIN APPLEBEE 17B512 OR EQUIVALENT
L6 12 TURNS NO. 30 ENAM. WIRE WOUND OVER
"COLD" END OF L5

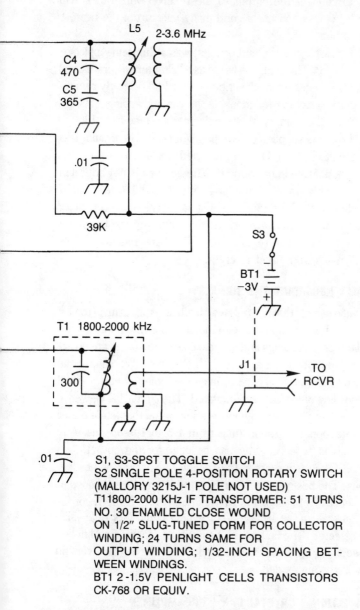

Fig. 6-13. Schematic diagram of the BC and LF converter. Except for C3, which is mica, capacitors of decimal value are disc ceramic. All others are mica or NPO ceramic. Resistors are 1/4 watt, or greater, composition.

100 to 200 kHz region. Do not try to use an unshielded i-f transformer or to connect the output to the receiver with other than coaxial cable, unless you are a good many miles from the nearest broadcast station.

The i-f transformer the author used was a Burstein-Applebee stock No. 18B120. This unit was listed as a "slug-tuned ceramic coil form," but it has windings as described under Fig. 6-13. By removing the original capacitors found inside the can and adding a 300 pF capacitor across the primary, the transformer will tune from 1800 to 2000 kHz. Power is supplied by two penlight cells. The drain is less than half a milliampere. The author tried additional voltage but without any noticeable improvement. The antenna is not critical in the 1600 to 200 kHz range, but a long wire at least 120 feet long is recommended for the VLF range. If built as specified this converter will provide plenty of performance all the way from the broadcast band to the never-never regions of VLF where radiated waves measure 17 miles from crest to crest.

GRAPHICALLY RECORDING THE SIGNALS

The radio signals that are received from any spectrum in the LF and VLF range have in the past been monitored in many ways. High speed audio tape recordings and graphic chart recordings are possibly the best methods used to record VLF signals. We choose the chart recording method because it gives the experimenter a visual graphic recording of the data accumulated. The circuit, as viewed in Fig. 6-14, shows an input-output integration circuit that lets the user vary the detection integration time from 1.5 to 10 seconds. Any standard millivolt type recorder can be used for the output. The input can be taken from either the receiver's speaker or headset connections. The best way to determine the optimum time constant for any given LF/VLF signal is by trial and error. If the recorder drives to full scale in the 10 second time period, a lesser time value should be selected. The unit can be housed in an aluminum Mini-box as can be seen in Fig. 6-15. Shielded cable between the receiver and integrator is recommended.

THE INTERNATIONAL CRYSTAL BAX-1 PREAMPLIFIER

Although the availability of preamplifiers for LF and VLF work is rather rare, International Crystal manufactures a BAX-1 Broad-

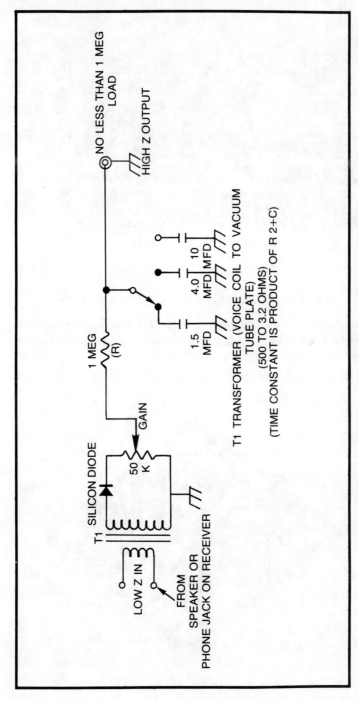

Fig. 6-14. Schematic of the graphic time constant integrator.

NO LESS THAN 1 MEG LOAD

HIGH Z OUTPUT

T1 TRANSFORMER (VOICE COIL TO VACUUM TUBE PLATE)
(500 TO 3.2 OHMS)
(TIME CONSTANT IS PRODUCT OF R 2+C)

10 MFD

4.0 MFD

1.5 MFD

1 MEG (R)

GAIN

50 K

T1 SILICON DIODE

LOW Z IN

FROM SPEAKER OR PHONE JACK ON RECEIVER

Fig. 6-15. An aluminum Mini-box for housing the integrator.

band Amplifier with a range from 20 Hz to 150 MHz. The amplifier may be used as either a tuned or untuned amplifier in rf or audio applications. For example, when used as an untuned rf preamplifier it is connected between the antenna and the receiver antenna posts. It

Fig. 6-16. The BAX-1 is manufactured as a kit.

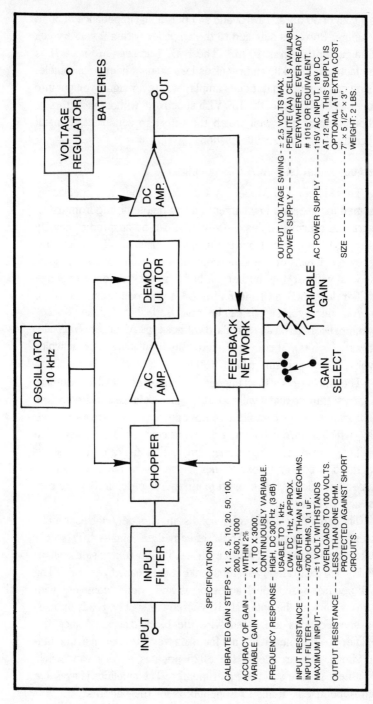

SPECIFICATIONS

CALIBRATED GAIN STEPS - X 1, 2, 5, 10, 20, 50, 100, 200, 500, 1000
ACCURACY OF GAIN - - - - - WITHIN 2%
VARIABLE GAIN - - - - - - - X 1 TO X 3000, CONTINUOUSLY VARIABLE.
FREQUENCY RESPONSE - HIGH, DC 300 Hz (3 dB) USABLE TO 1 kHz.
 LOW, DC 1Hz, APPROX.
INPUT RESISTANCE - - - - - GREATER THAN 5 MEGOHMS.
INPUT FILTER - - - - - - - 4700 OHMS, 0.1 uF.
MAXIMUM INPUT- - - - - - ±1 VOLT. WITHSTANDS OVERLOADS TO 100 VOLTS.
OUTPUT RESISTANCE - - - - LESS THAN ONE OHM. PROTECTED AGAINST SHORT CIRCUITS.

OUTPUT VOLTAGE SWING - ± 2.5 VOLTS MAX.
POWER SUPPLY - - - - - PENLITE (AA) CELLS AVAILABLE EVERYWHERE, EVER READY #1015 OR EQUIVALENT.
AC POWER SUPPLY - - - - -115V AC INPUT. 18V DC AT 12 mA. THIS SUPPLY IS OPTIONAL AT EXTRA COST.
SIZE - - - - - - - - - -7" × 5 1/2" × 3".
 WEIGHT: 2 LBS.

Fig. 6-17. Block diagram of the Abbeon Cal LA-1 ac/dc preamplifier.

has a gain of 30 dB at 1 MHz and an operational impedance of 50 to 500 ohms. Power is supplied to the amplifier with a 9 volt dry cell with a current drain of 10 mA. The BAX-1 as seen in Fig. 6-16 is manufactured as a kit, and requires less than an hour to assemble. When you consider that it does amplify whatever signal you put into it, over a goodly part of the LF/VLF spectrum, without inductors or without tuning, it is well worth the effort to construct. Consult Appendix B for further information.

The Abbeon Cal LA-1 AC/DC Preamplifier

The Abbeon Cal Model LA-1 ac/dc amplifier was specifically designed for use with low current and voltage emitting from transducers and thermocouples; however, it can be adapted to amplify and record the output from the speaker terminals of any low frequency receiver. For the experimenter who wants to have graphic readouts without the bother of building the current-to-voltage amplifier described in Chapter 4 the LA-1 is a good substitute. With the LA-1, microvolt signals can be measured, shown on any oscilloscope, or recorded on any standard voltage type recorder. When set to a gain of ×1000, it converts microvolts to millivolts, or millivolts to volts. The maximum gain is ×3000.

The amplifier is also an impedance converter, which operates at any gain setting so that it will accept a high resistance sensor, such as a photodiode, yet give sufficient output power to operate a sensitive relay, which in turn can trigger an event counter or an alarm. The amplifier, shown as a block diagram in Fig. 6-17, is not just an integrated circuit in a box. Rather it uses ultrastable circuits developed specifically for measurement and amplification of low current and voltage.

The circuit includes both a solid state chopper/stabilizer and a battery voltage regulator. These effectively eliminate drift, so the LA-1 has no front panel zeroing or setting-up controls. Excellent stability and calibrated gain steps make it very easy to operate. The LA-1 needs no additional accessories and comes complete with batteries, or with the optional 115 Vac power supply, ready for use. The specifications are listed with the block diagram. Since the amplifier is rather new it has not found its way into used instrumentation shops. Nevertheless, the $225 price is a very worthwhile investment for the serious experimenter. The amplifier is available from Abbeon Cal. Refer to Appendix B for their address.

Chapter 7
Batteries And Power Supplies

Solar flare recording involves long term monitoring of atmospheric conditions with equipment susceptible to interference from power line hum and other manmade causes. These conditions make operation from batteries particularly attractive, especially when equipment is to be installed at remote locations where unattended operation prevents monitoring of power failures. Whether to reduce interference, to ensure reliable power, or to secure both conditions, an experimenter should pay close attention to the source of power for the station.

GEL/CELL BATTERIES

The use of rechargeable gel/cell batteries as portable power supplies for electronic use has become increasingly popular, especially in amateur transceivers and other electronic gear, due to a number of reasons. First, they offer portability to the electronic gear; second, they last five times longer than the conventional carbon or alkaline batteries; third, they have a very long shelf and operating life; and fourth, they can withstand a wide temperature operating range. On some occasions it may be desirable to house the solar flare receiver in a remote area that has no ac power source. In such a case the need for a dependable dc voltage source is fulfilled by using a gel/cell type battery. The use of carbon or alkaline batteries in remote, unattended operations could cause problems since they are prone to temperature changes, leakage, and short operating and shelf life. The gel/cell battery contains a jelled electrolyte. There is no liquid to splash around or ooze out. Internally, the battery is drier than a dry cell and has most of the advantages of the wet cell. It can be recharged again and again. If the battery is not completely discharged during each cycle, 1000 or more cycles of operation are possible. When *floated* at a constant voltage of 2.25 to 2.30 volts per cell, the gel/cell can be maintained for *years*. The following is a summary of the operating conditions and recharging capabilities of the gel/cell.

Recharging and General Conditions

The open circuit voltage of each cell of the gel/cell battery is approximately 2.12 volts. The cell voltage is sometimes lower for a severely discharged battery and sometimes higher for a battery that has just been taken off charge, but in all instances it should adjust to 2.12 volts after a period of time. To recharge the gel/cell battery, a dc voltage, greater than the open circuit voltage of the battery, is applied to the battery terminals. The plus terminal of the charger should be connected to the positive (red) terminal of the battery and the negative terminal of the charger to the negative (gray) terminal of the battery. Limit the initial charge current that flows into the battery to the values shown in the Table 7-1.

As the battery begins to accept charge, its voltage will rise. Normal end-of-charge voltage is 2.4 volts per cell (measured while the charge current is flowing). A 6 volt battery, for example, should

Table 7-1. Maximum Initial Charge Current for Typical Gel/Cell Batteries.

Battery Rating (Ampere Hour)	Initial Charge Current (Amps)
.9	0.15
*1.5	0.25
1.8	0.30
2.6	0.4
4.5	0.7
6.0	0.9
7.5	1.2
20.0	4.0

*Globe No. GC-1215-1 1.5 ampere hour, 12 volt is recommended type for flare receiver use.

be charged to a voltage of 7.2 volts and a 12 volt battery should be charged to a voltage of 14.4 volts. Voltage is only one of *two* indicators that must be used to determine if a battery is fully charged. The other is current. A battery is not fully charged until the current, at 2.4 volts per cell, drops to the following value: for the gel/cell used with solar flare receivers (1.5AH; 0.25A) this should be 20-40 mA, which is the approximate final current. When the current into the battery, at 2.4 volts per cell, reaches this value, the charger should be disconnected from the battery. This will prevent the battery from being overcharged and will insure optimum battery performance and life.

"Continuous Charge"

If it is desired to maintain the gel/cell battery "on charge" continuously and to allow the battery charger to remain plugged into the ac power line without ever disconnecting it, the charge voltage should be held at 2.25 volts per cell, *not* at 2.4 volts. If held at 2.4 volts, battery life will be significantly reduced because of over-charge. When held at the recommended float voltage of 2.25 volts per cell, the battery will seek its own current level to maintain itself in a fully charged condition. It will also fully recharge itself after a power outage, but, of course, not quite as rapidly as when a higher voltage is used. There may be some reduction in the total number of charge/discharge cycles when compared to using a higher charge voltage, but the total length of battery life will not be affected. The 2.25 volts per cell is considered a float voltage. The term *float voltage* should not be confused with the term *trickle charge*. Float voltage refers to the condition where the charging voltage is held constant and the charging current is free to vary. In contrast, trickle charge refers to the condition where the charging current is held constant and the voltage is free to vary. When maintaining the gel/cell battery it is recommended that a float voltage be used in preference to a trickle charge. It is felt the trickle charge method is more apt to overcharge the battery.

Cyclic Use

It is important to obtain the maximum number of recharge cycles; thus, the battery on-charge voltage should be first brought up to 2.4 volts per cell and held there until the current drops to

20-40 mA (for the GC-1215-1). The charger should then be switched to a float voltage of 2.25 volts per cell or switched off.

Battery Life

Using the "cyclic use" method of recharge, 200 to 400 or more complete full charge/full discharge cycles are possible. Of course, if the battery is only slightly discharged during each cycle, instead of being totally discharged, literally thousands of cycles of operation are possible. The author has been using two Globe gel/cell batteries at a remote solar flare station for over two years. One supplies power to the LF solid state receiver, the other powers an inverter that supplies power to the 110 volt recorder. Over 50 full charge/full discharge cycles have been logged on the batteries.

Ratings

Gel/cell batteries are rated at the 20 hour rate. This means a 2.6 ampere hour battery, for example, will put out 0.130 amps for 20 hours (20 hours × 0.130 amperes = 2.6 ampere hours). This does not mean, however, that a 2.6 ampere hour battery will put out 2.6 amperes for 1 hour. A 2.6 ampere hour battery will actually put out 1.7 amperes for 1 hour. Therefore, the capacity of a 2.6 ampere hour battery at the 1 hour rate would be 1.7 ampere hour. It is typical for all battery systems that as the rate of discharge increases, the total capacity available from the battery decreases. A new gel/cell battery has an initial capacity of 80 to 90 percent of its nominal rating. After several months of storage, or after 30 to 40 complete charge cycles, the nominal capacity is reached. The capacity then remains at the nominal value up to 150–200 cycles. An increase in the initial capacity of the battery can be obtained through several low rate (20 hours or longer) discharges before use. Partial discharges or discharges at a high rate of current will also accelerate the increase in capacity, but more slowly.

BATTERY CHARGERS

From the previous discussion, battery chargers obviously must be selected with care. At the least, battery life can be reduced, at the worst, ruined.

Transformer/Rectifier

There are many types of chargers that can be used on the gel/cell battery. The simplest and lowest in cost is a small transformer/rectifier circuit which has a secondary of the transformer wound so that the current drops when the battery on-charge voltage rises. Figure 7-1 shows the basic circuit. A ratio of 9 or 10 to 1 between start and finish currents is about the maximum for this type of charger. For example, if the charger is designed for a final charge current of 40 mA at 2.4 volts per cell, the highest initial current obtainable would be about 400 mA. This type of charger has the disadvantage of being quite sensitive to line voltage variations.

Series Transistor

Another type of charger is illustrated in Fig. 7-2. It is basically a simple series transistor regulator with a zener diode reference. Many variations of this type circuit can be found in any transistor handbook. This type circuit does a good job compensating for live voltage variations. Because of the amplification effect of the transistor, the ratio of start to finish current can be 20 to 1 or higher. This permits a higher start-current without getting a correspondingly higher than normal end-of-charge current. The higher start-current means faster recharging.

Shunt Zener

In lieu of a transistor and small zener as in the series transistor circuit, it is possible to use just a shunt connected zener as is shown

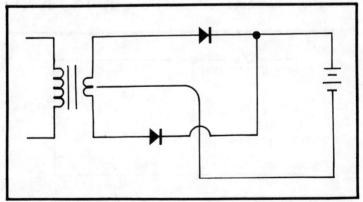

Fig. 7-1. Schematic diagram of a transformer/rectifier type battery charger.

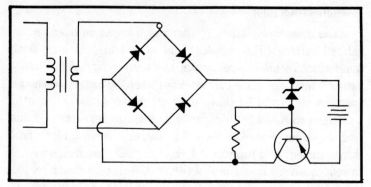

Fig. 7-2. Schematic diagram of a series transistor type battery charger.

in Fig. 7-3. This circuit is very economical for small capacity batteries because inexpensive, low-wattage zeners can be used.

Shunt Transistor

Another circuit often used is the shunt transistor regulator as shown in Fig. 7-4. This circuit permits a small zener to be used for all battery sizes since only the transistor base current flows through the zener.

Application Notes

Remember that whenever an end-of-charge voltage greater than 2.25 volts is used, the charger should be disconnected from the battery at the end of charge. This will keep the battery from being overcharged and will insure optimum battery life. Gel/cell batteries should not be stored in a discharged condition, and they should be

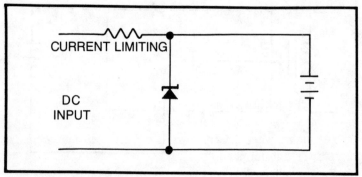

CURRENT LIMITING

DC
INPUT

Fig. 7-3. Schematic diagram of a shunt zener type battery charger.

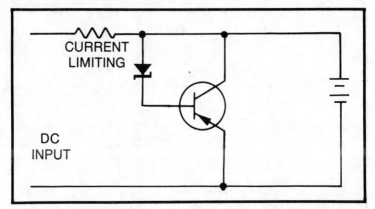

Fig. 7-4. Schematic diagram of a shunt transistor type battery charger.

recharged as soon as possible after each use. Batteries that are not recharged soon after discharge, or are stored in a discharged state, may appear to be open circuited when an attempt is made to recharge them. Or else they will accept far less current than normal. When this condition is encountered, just leave the charger connected to the battery. After a period of time the battery will begin to accept normal current, or will accept larger and larger amounts of current until the normal current level is reached. On future recharges, the battery will behave in a normal manner unless it is again stored in a discharged state. Gel/cell batteries are a product developed by Globe Battery, Division of Globe-Union, Inc. They are available at major electronic and electrical supply houses. A distributor listing in your area can be obtained by writing to Globe at the address listed in Appendix B.

A HOME BREW POWER SUPPLY

The power supply that we are about to describe has a three fold purpose. First, it is easily constructed from salvage parts (an old TV set transformer). Second, it can be used to power any number of 12 volt solid state devices including any of the LF-VLF receivers described in this chapter, and third, the cost of the entire supply is less than ten dollars. The problem the amateur/experimenter usually encounters when building a power supply is trying to find the right power transformer to provide the necessary current at the appropriate voltage (and cost). One way to cut costs would be to rewind a power transformer from an old TV set; however, before

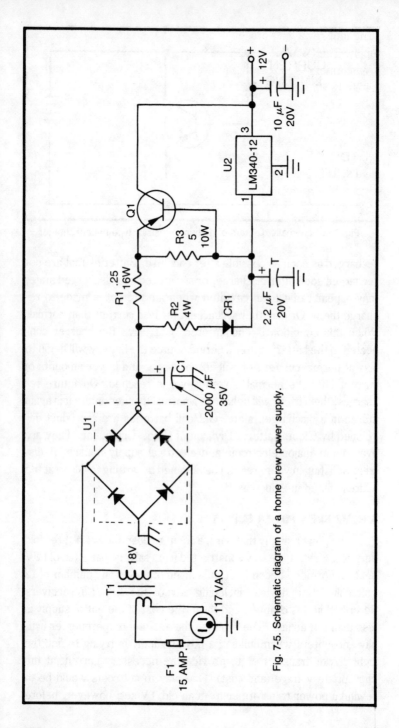

Fig. 7-5. Schematic diagram of a home brew power supply.

going into rewinding procedures, let us briefly discuss the power supply. The circuit for the home brew supply is shown in Fig. 7-5. Transformer T1, a rewound TV transformer, provides approximately 18 volts ac, which is rectified via U1 and then regulated at 12 volts. The regulator used here is a National Semiconductor LM340K-12. A PNP pass transistor is used to increase the current draw of the power supply. This home brew supply is capable of delivering up to 10 amperes to power any 12 volt electronic gear and *more* than capable of powering any of the LF-VLF receivers described in this chapter.

REWINDING POWER TRANSFORMERS

The amateur/experimenter may feel that rewinding a power transformer is a specialized task. However, you may be reassured it is a relatively simple process and the only expenditure is time. Old TV sets are an excellent source of supply for the transformers. For the type of power supply we are going to describe here, practically any TV transformer has the necessary power capabilities. Twelve volts at 10 amperes = 120 watts, and even with a 100 percent excess rating (240 watts) such transformers are common in TV sets. Before getting into the actual description of rewinding a transformer, let's take a closer look at a typical power transformer to see how it is constructed. A power transformer consists of a laminated iron core, windings of various sizes to provide the necessary voltages and currents, insulating paper, nuts and bolts to hold the unit together, and metal covers to protect the windings. The iron laminations consist of E- and I-shaped sections as shown in Fig. 7-6. In the actual construction of a transformer the laminations may be put together in groups. In other words, there may be three E and I sections stacked the same way, then three more of each type section stacked in the alternate arrangement. An insulating or bonding medium, usually varnish or shellac, is applied between laminations. The purpose of this is to reduce the poer loss in the core, and also to serve as a bonding for a tight form, minimizing hum or vibration.

Determining Power Capabilities

One of the first things a builder must know when acquiring an old TV power transformer is how much power it will handle. If you are going to build a power supply that requires 300 watts of power,

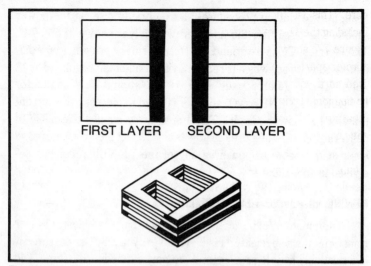

Fig. 7-6. How the core is assembled. Alternate layers have E laminations facing oppositely. Occasionally two or more laminations of the same kind are grouped together and handled as a single lamination to save cost and assembly time.

you cannot get it from a transformer that has only 200 watts capability. The amount of power that a transformer will handle can be determined quite accurately from the cross-sectional area of the

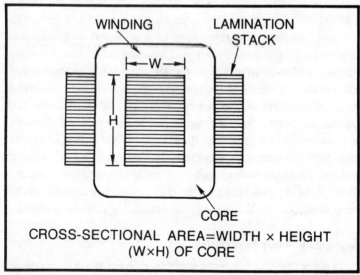

Fig. 7-7. A cross sectional drawing of a power transformer. The cross sectional area referred to in Fig. 7-8 is determined by multiplying the tongue width by the height of the center core.

core. This is, the cross sectional area *inside* the windings, not including the area of the part of the core that surrounds the winding. Figure 7-7 shows this area. It is not necessary to take the transformer apart to measure this area. Lamination sizes are standardized so if you know the outside width, length, and height of the lamination stack it is easy to determine the power capabilities. Nearly all TV transformer cores have the same width and length, but the height of the stack will vary. The width and length are commonly 3-3/4 by 4-1/2 inches, and for a core of this size the tongue of the E lamination is always 1-1/2 inches wide. With such a core all that is needed for finding the cross sectional area is the height of the stack. For example, suppose the height is 2-1/4 inches. This, multiplied by 1-1/2, equals 3-3/8 square inches. Looking at the graph in Fig. 7-8 we can see that 3-3/8 square inches gives a power capability of 350 watts. This means that we can rewind a transformer having a core of these dimensions and get about 350 watts from it. There are several other facts to consider when looking for an old TV transformer. First, take one with the highest stack of laminations. This will be the one with the best power capabilities. Second, some transformer manufacturers soak the coils in tar. This type can be re-wound but it can be a rather messy job and is best avoided.

Taking Apart Transformer

The first step in the rebuilding process is to remove the transformer from the TV chassis. You can save yourself some further work if you first check out the windings and label them. The primary or input winding will be connected to the ac line, probably through a switch on the front of the chassis and a fuse or fuse holder on the rear. The 5 volt winding will be connected to the filament terminals (2 and 8) on the rectifier socket, which is usually a 5U4G. Two of the leads from the high voltage winding will be connected to the plate terminals (4 and 6) on the rectifier tube socket. The center tap lead of the high voltage winding will be grounded to the chassis. There will probably be two 6.3 volt windings. The leads from one of these will go to the shielded compartment on top of the TV chassis and be connected to a tube socket in the compartment. The other 6.3 volt winding supplies are the other tube heaters in the TV set. Tag all leads before removing the transformer.

If you acquire a transformer that has *already* been removed from a TV set, it is best to check the various windings with an ac

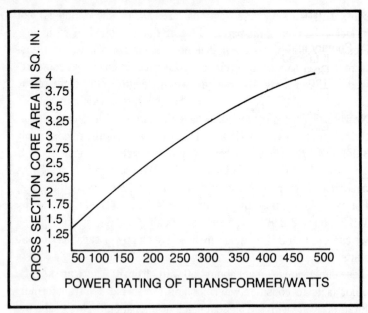

Fig. 7-8. To determine power capabilities of an unknown transformer, first determine the cross sectional area of the core. Locate this area value on the vertical axis and follow it across until intersecting the curve. From this point drop straight down to the horizontal axis for the power rating in watts.

voltmeter and mark them. There is a color code used for the transformer leads and the information for identifying the leads can be seen in Table 7-2; however, the leads are not always marked according to the code. Also, the colors tend to fade with age, so it is best to actually check the transformer with an ac voltmeter. When checking out the transformer with a meter you may find that voltages are slightly higher than what is actually called for because of a no-load condition on the windings. For example, the 5 volt rectifier winding will show something over 5 volts, but if the leads come off the rectifier socket, the winding is a 5 volt winding.

After identifying the windings, remove the four nuts and bolts that hold the transformer together and also take off the metal covers, assuming the unit has them. Look the unit over carefully and try to determine which layers of windings are which. In most cases the winding nearest the core will be the primary. Usually the order will be something like this: First, the primary; next, the high voltage; then, the 5 and 6.3 volt, the heavier-current 6.3 volt winding. Examine the lamination arrangement. Note that the laminations are

Table 7-2. Color Codes Used With Standard Power Transformers.

Primary leads _ _ _ _ _ _ _ _ _ _ _ black
   if tapped:
      Common _ _ _ _ _ _ _ _ _ _ _ black
      Tap _ _ _ _ _ _ _ _ _ _ _ _ _ black & yellow striped
      Finish _ _ _ _ _ _ _ _ _ _ _ _ black & red striped

High voltage plate winding _ _ _ _ _ _ red
      Center-tap _ _ _ _ _ _ _ _ _ _ _ red & yellow striped

Rectifier filament winding _ _ _ _ _ _ yellow
      Center-tap _ _ _ _ _ _ _ _ _ _ _ yellow & blue striped

Filament winding No. 1 _ _ _ _ _ _ _ green
      Center-tap _ _ _ _ _ _ _ _ _ _ _ _ green & yellow striped

Filament winding No. 2 _ _ _ _ _ _ _ brown
      Center-tap _ _ _ _ _ _ _ _ _ _ _ _ brown & yellow striped

Filament winding No. 3 _ _ _ _ _ _ _ slate
      Center-tap _ _ _ _ _ _ _ _ _ _ _ _ slate & yellow striped

probably inserted in groups. On one side of the stack there may be three *I* units and below that three *E* units, alternating through the entire stack. Note how the top and bottom of the stack are assembled so that you will be able to put it back in this same manner when you complete the winding job.

Getting the laminations apart is not a difficult job, but it should be done carefully. Insert a thin knife blade between the end piece and the rest of the core to break the varnish seal so that the end piece will be loose. Using a block of wood batted against the edge of the piece, drive it out of the core with light taps of the hammer. It is a good practice to alternate between the two ends so that the piece will come out straight. Continue by breaking the next group of laminations free with the knife blade, then carefully driving them out. After a few groups have been removed, the hammer won't be needed, as the broken-loose laminations can be pulled out by hand. Be careful not to bend the laminations when removing them. If the edges get nicked by the hammering, file them smooth before reassembling the core after the new windings are finished.

Once the laminations are removed you are ready to go to work on the windings. The first thing to do is remove the high voltage winding by pulling out the wire. With luck, you can start to do this by pulling on one of the high voltage leads. However, it is more than likely that the end of the winding will break off because the wire size will be rather small. If it breaks you will have to dig in with a knife or probe to get at the wire. Once you get it started the layers come out rather easily. With most of the high voltage winding it should become apparent that you can separate the primary winding section from the outer windings. Be careful not to disturb the insulation around the primary winding. The wire size on most TV transformers used in the primary is No. 18 enameled. After you have cleared away the high voltage winding, remove the 5 volt rectifier-filament winding and *most* carefully count the number of turns. There will be approximately 10 turns but count them to make sure. The number of turns on this winding will tell you how many turns you need for each volt you expect to get with the new windings you will wind on. For example, if there are 10 turns on the 5 volt winding, the transformer is wound on the basis of two turns per volt. It doesn't make any difference whether the windings are near the center of the core or at the side; the turns per volt will be the same.

Putting on the New Winding

For this home brew power supply, the current rating is to be 10 amperes, so a wire size that will carry the current is required. No. 12 solid enamel-covered wire handles this requirement. The transformer used in the author's power supply required two turns per volt, and 18 volts ac was needed. This works out to a total of 36 turns of No. 12 wire. To calculate how much wire you need, take a scrap length of wire or string, make a full turn around the core containing the primary winding, then measure how long the piece of wire or string is. Multiply this by 38, then add about three feet for lead lengths. Clamp one end of the good wire in a vise and, making sure there are no kinks, start winding the wire over the section that had the primary winding. Start as close to the edge as possible and keep the wire taut as you wind on the turns. The reason for starting close to the edge is that as you put layers on, each layer has to be progressively narrower; otherwise the end turns may slip off. After the first layer is wound, hold the ends in place with Scotch tape.

Ordinary household wax paper can be used between layers. A single layer or sheet of paper is adequate between layers (for convenience a supply of cut up pieces of wax paper should be prepared prior to winding). Wrap a sheet tightly around the first layer of the winding and fasten the end of the paper with small pieces of Scotch tape. Try to keep the starting point for the next layer as close to the outside turn of the previous layer as possible and always wind in the *same* direction. Take your time, keeping the wire taut, and by all means, keep track of your count by *making notes*. It is easy to lose your place in counting. Be sure to bring all leads and taps out on the same side of the core so the transformer covers will go back in place without any problems. Note how it was done originally, before taking the transformer apart. Once all of the turns are on, cover the windings with several layers of electrician's tape. The transformer can now be put back together. If there is too much open area between the top and bottom of the windings and the iron core, make up some smooth wooden wedges and gently drive them between the windings and the core. This will help prevent transformer hum or rattle. Tubing insulation can be slipped over the leads where they come through the transformer housing to prevent chafing of the wire enamel covering.

Additional Details

To keep the cost as low as possible, one needs to search the surplus sales sheets. Pass transistor Q1 that the author used was strictly unbranded surplus. It came with a heat sink that was large enough to make a 10 ampere rating seem reasonable. Any PNP type that has 10 amperes or more of collector-current rating is adequate. Some suitable Motorola HEP numbers are 233 and 237. Under their new numbering system, these would be G6006 and G6010, respectively. CR1 can be any rectifier that has a rating of at least 3 amperes at 35 volts. Surplus got into the act again in the author's prototype in obtaining the required resistor values for R1, R2, and R3. Resistor R1 was managed by putting together four 5 watt, 1 ohm resistors in parallel. A heat sink is recommended for Q1. Usually, the case of the power PNP transistor is also the collector connection, and the case must be in contact with the heat sink. If so, this means that the heat sink should be "floated" or insulated from the main power supply chassis. Although this home brew power supply is made mostly of surplus and scrap parts it is equivalent to power supplies costing $100 or more on the commercial market. It should give the amateur/experimenter a healthy challenge to wind his own transformer and finish with a first class regulated unit.

Appendix A
VLF Stations

kHz	Call	Location	Power (kW)	Remarks
14.1	NAA	Cutler, ME	1 MegW	to Peking, P.R.C.
14.29	SOA20	Warsaw, Poland	200	
14.3	UBE2	Petropavlovsk Kam., USSR	500	
14.5	HWU	Le Blanc, France	250	into North Atlantic Ocean
14.6	UVA	Batumi, USSR	100	into Black Sea
14.7	NAA	Cutler, ME	1 MegW	
	NEJ	Seattle, WA	1 MegW	worldwide
	NHB	Kodiak, AK	1 MegW	into N. Pacific Ocean
	NLK	Seattle, WA	1 MegW	
	NPM	Pearl Harbor, HI	1 MegW	
	NPN	Guam	1 MegW	worldwide
14.8	NAA	Cutler, ME	1 MegW	
14.881		Komsomol'sk na Amure, USSR	500	Navigational system
		Krasnodar, USSR	500	Navigational system
		Novosibirsk, USSR	500	Navigational system
14.9	NBA	Balboa, Canal Zone	1 MegW	worldwide
15.0	UIK	Vladivostok, USSR	100	Sea of Japan
15.1	FUO	Bordeaux, France	500	worldwide
	HWU	Le Blanc, France	500	worldwide
	VII	Bombay, India	100	Arabian Sea/Indian Ocean
15.3	NEJ	Seattle, WA	1 MegW	
	NHB	Kodiak, AK	1 MegW	
	NLK	Seattle, WA	1 MegW	
	NPM	Pearl Harbor, HI	1 MegW	
	NPN	Guam	1 MegW	
	EVT2	Ostrov, USSR	200	Arctic Ocean
15.5	NAA	Cutler, ME	1 MegW	worldwide
	NPM	Pearl Harbor, HI	1 MegW	worldwide
	NSS	Washington, DC	1 MegW	
	NWC	Northwest Cape, Australia	1 MegW	
15.6	EWB	Odessa. USSR	1 MegW	worldwide

kHz	Call	Location	Power (kW)	Remarks
15.625		Komsomol'sk na Amure, USSR	500	Navigational system
		Krasnodar, USSR	500	Navigational system
		Novosibirsk, USSR	500	Navigational system
15.7	NPG	San Francisco, CA	500	net
	NPL	San Diego, CA	500	net
	NPM	Pearl Harbor, HI	1 MegW	net
15.975	GBR	Rugby, England	750	worldwide
16.0	GBR	Criggion, England	250	Time/Standard
16.2		Yosami, Japan	500	worldwide
	UGK	Kaliningrad, USSR	500	worldwide
16.4	JXN	Helgeland, Norway	350	
16.5	SOA30	Warsaw, Poland	200	to Peking, P.R.C.
16.6	NPM	Pearl Harbor, HI	1 MegW	
16.8	FTA2	St. Assise, France	250	
17.1	UMS	Moscow, USSR	1 MegW	
17.2	SAQ	Varberg, Sweden	200	
17.44		Yosami, Japan	500	worldwide
17.6	JXZ	Helgeland, Norway	350	
	SAO40	Warsaw, Poland	200	
17.8	NAA	Cutler, ME	1 MegW	
	NPM	Pearl Harbor, HI	1 MegW	
	NSS	Washington, DC	1 MegW	
17.9	UBE2	Petropavlo, USSR	500	worldwide
18.0	NBA	Balboa, Canal Zone	500	
	NLK	Seattle, WA	1 MegW	
	NPG	San Francisco, CA	1 MegW	
	NPL	San Diego, CA	1 MegW	
18.1	UFOE	Matochkin Shar, USSR	100	Sea of Norway
18.2	JJH	Kure, Japan	500	
	NAH	New York, NY	200	
	NSS	Washington, DC	1 MegW	
18.4	NAD	Boston, MA	200	
	NAH	New York, NY	200	
	NAK	Annapolis, MD	200	
18.5	NAA	Cutler, ME	1 MegW	
18.6	NAA	Cutler, ME	1 MegW	
	NEJ	Seattle, WA	1 MegW	
	NHB	Kodiak, AK	1 MegW	
	NLK	Seattle, WA	1 MegW	
	NPG	San Francisco, CA	1 MegW	
	NPM	Pearl Harbor, HI	1 MegW	worldwide
	NPN	Guam	1 MegW	worldwide
18.8	NAD	Boston, MA	200	
	NAH	New York, NY	200	
	NAK	Annapolis, MD	200	
18.9	UMV	Rostov, USSR	1 MegW	worldwide
19.0	GQD	Anthorn, England	500	
	MHW	Criggion, England	350	
	MHW	Rugby, England	250	
	NPM	Pearl Harbor, HI	1 MegW	
	NSS	Washington, DC	1 MegW	to Peking, P.R.C.
19.2	SOA50	Warsaw, Poland	200	worldwide
19.4	NEJ	Seattle, WA	1 MegW	worldwide
	NHB	Kodiak, AK	1 MegW	
	NLK	Seattle, WA	1 MegW	

kHz	Call	Location	Power (kW)	Remarks
	NPM	Pearl Harbor, HI	1 MegW	
	NPN	Guam	1 MegW	
19.6	GBZ	Anthorn, England	500	worldwide
	GBZ	Criggion, England	350	
	GBZ	Rugby, England	250	worldwide
19.7	UGE	Arkhanghelsk, USSR	150	into Barents Sea
19.8	NPG	San Francisco, CA	1 MegW	net
	NPL	San Diego, CA	1 MegW	net
	NPM	Pearl Harbor, HI	1 MegW	worldwide
20.0	JG2AR	Tokyo, Japan	3	standard frequency
	WWVL	Boulder, CO	40	standard frequency
20.27	ICV	Tavolara, Italy	500	worldwide
20.76	ICV	Tavolara, Italy	500	worldwide
21.05	HWU	Le Blanc, France	200	North Atlantic Ocean
21.37	GYA	London, England	120	North Atlantic Ocean
21.4	NAA	Cutler, ME	1 MegW	
	NPM	Pearl Harbor, HI	1 MegW	
	NSS	Washington, DC	1 MegW	
21.75	HWU	Le Blanc, France	100	North Atlantic Ocean
22.1	NAD	Boston, MA	200	
	NAH	New York, NY	200	
	NAK	Annapolis, MD	200	
22.3	NAA	Cutler, ME	1 MegW	
	NPC	Seattle, WA	1 MegW	
	NPM	Pearl Harbor, HI	1 MegW	
	NWC	Northwest Cape, Australia	1 MegW	
22.35	NAA	Cutler, ME	1 MegW	Atlantic Ocean
	NAD	Boston, MA	200	
	NAH	New York, NY	250	
	NAK	Annapolis, MD	200	
23.0	UFQE	Matochkin Shar, USSR	100	into Barents Sea
23.4	DMB	Mainflingen, G.F.R.	10	to New York
	NPM	Pearl Harbor, HI	1 MegW	
24.0	NBA	Balboa, Canal Zone	1 MegW	worldwide
	NLK	Seattle, WA	1 MegW	
	NPM	Pearl Harbor, HI	1 MegW	
25.3	NAA	Cutler, ME	1 MegW	North Atlantic Ocean
25.82	NAA	Cutler, ME	250	net
	NAH	New York, NY	200	North Atlantic Ocean
	NSS	Washington, DC	1 MegW	North Atlantic Ocean
26.1	NEJ	Seattle, WA	1 MegW	
	NPG	San Francisco, CA	1 MegW	
	NPM	Pearl Harbor, HI	1 MegW	net
26.6	CAA2A	Santiago, Chile	50	net
27.0	FTA27	Paris, France	25	to New York
27.5	EWC4	Vladivostok, USSR	80	into Sea of Japan
28.5	NAU	Fort Allen, PR	100	Atlantic Ocean
	NPL	San Diego, CA	500	net
	NPN	Guam	50	net
28.64	RAM	Moscow, USSR	1 MegW	worldwide
30.0	PWI	Recife, Brazil	20	
30.3	UGK2	Kaliningrad, USSR	100	into Baltic Sea
30.6	NAF	Newport, RI	50	net
	NPC	Bainbridge, WA	100	net
	NPG	San Francisco, CA	50	net

kHz	Call	Location	Power (kW)	Remarks
	NPL	San Diego, CA	500	net
	NUD	Adak, AK	100	net
32.55	FTA32	St. Assise, France	25	
32.9	NAU	Isabella Segun, PR	100	Atlantic Ocean
	NBA	Balboa, Canal Zone	50	net
33.3	NGR	Kato Soli, Greece	100	Mediterranean Sea
33.95	LCD	Jeloey, Norway	25	net
36.55	LCD	Jeloey, Norway	25	net
37.0	UWR	Murmansk, USSR	50	Arctic Ocean
38.0	TFK	Reykjavik, Iceland	50	Washington, DC
40.0	JJF2	Tokyo, Japan	30	standard frequency
40.4	SAS	Varberg, Sweden	40	
40.75	GXH	Thurso, Scotland	100	North Atlantic Ocean
	NAM	Norfolk, VA	100	Guantanamo
	NAU	Isabella Segun, PR	100	
	NPG	San Francisco, CA	100	
	NPL	San Diego, CA	100	
	NPN	Guam	100	
	NPO	Luzon, Phillipines	50	
	NSS	Washington, DC	5	
	NST	Londonderry, N. Ireland	50	New York
41.5	ORL72	Ruiselede, Belgium	40	Indian Ocean
42.0	ARL	Karachi, Pakistan	40	
42.55	SAS2	Varberg, Sweden	40	traffic
43.2	GIX20	Oxford, England	60	Washington, DC
43.79	RAM	Moscow, USSR	120	
44.0	VHB	Belconnen, Australia	200	
	VIX	Belconnen, Australia	200	
44.25	SAS3	Varberg, Sweden	40	
44.95	GYN1	London, England	40	
	GYW	Gibraltar	40	
45.6	SAS4	Varberg, Sweden	40	
45.75	VFH	Halifax, NS, Canada	250	North Atlantic Ocean
46.25	DCF46	Mainflingen, G.F.R.	5	net
46.58	GYW	Gibraltar	40	
46.85	GYC	London, England	40	
	MTO21	Crimond, England	40	
47.45	DCF47	Mainflingen, G.F.R.	40	Baltic/North Sea
	DHJ54	Cuxhaven, G.F.R.	40	Baltic Sea
	DHJ57	Kiel, G.F.R.	40	Baltic Sea
	DHJ58	Flensberg, G.F.R.	40	Baltic Sea
	DHJ59	Wilhelmshaven, G.F.R.	40	Baltic Sea
	NAU	Fort Allen, PR	100	
	NHB	Kodiak, AK	50	
	NPL	San Diego, CA	50	
	NUD	Adak, AK	50	
48.05	CUC4	Lisbon, Portugal	40	
48.6	OLP	Prague, Czechoslovakia	25	Moscow
49.55	SAV	Karlsborg, Sweden	40	
50.0	OMA	Prague, Czechoslovakia	1	standard frequency
50.2	OLP3	Prague, Czechoslovakia	25	
50.45	OLP2	Prague, Czechoslovakia	25	Tirana/Bucharest
50.75	FTA50	Lyon, France	135	
51.25	ORL58	Ruiselede , Belgium	40	
51.6	NPL	San Diego, CA	100	

kHz	Call	Location	Power (kW)	Remarks
	NUD	Adak, AK	50	
51.7	EWC4	Vladivostok, USSR	30	into Sea of Japan
51.95	GIY20	Oxford, England	60	
	GYA	London, England	60	North Atlantic Ocean
52.3	RTO	Moscow, USSR	100	Berlin
52.65	NGR	Kato Soli, Greece	100	Indian Ocean
54.0	NDI	Naha, Okinawa	50	
	NDT2	Yokosuka, Japan	200	
	NPN	Guam	100	
54.05	NBA	Balboa, Canal Zone	50	
	NPL	San Diego, CA	50	
	NUD	Adak, AK	50	
54.65	LCH	Oslo, Norway	20	
55.25	DCF55	Mainflingen, G.F.R.	10	
55.5	CNL	Kenitra, Morocco	100	Mediterranean Sea
	GXH	Thurso, Scotland	100	
	NPG	San Francisco, CA	250	Pearl Harbor
	NPM	Pearl Harbor, HI	50	San Francisco
	NPO	Luzon, Phillippines	50	
55.75	SOA60	Radom, Poland	40	Helsinki
56.35	GLP20	Ongar, England	25	Madrid/Lisbon
57.1	OXE20	Skamlebaek, Denmark	20	
57.4	CNL	Kenitra, Morocco	50	
57.7	LBH	Trondheim, Norway	60	
57.9	NAU	Isabella Segun, PR	100	net
	NEJ	Seattle, WA	50	
	NPG	San Francisco, CA	50	
58.25	SOA70	Radom, Poland	360	Helsinki/Oslo
58.3	NGR	Kato Soli, Greece	100	Indian Ocean
59.0	NGR	Kato Soli, Greece	100	Mediterranean Sea
60.0	GBT20	Rugby, England	80	Rome/Lisbon/Madrid
	MSF	Rugby, England	80	standard frequency
	WWVB	Boulder, CO	3	standard frequency
61.75	GYA1	London, England	15	
	NHB	Kodiak, AK	50	
	NUD	Adak, AK	50	
62.45	SOA80	Radom, Poland	40	Helsinki/Olso
62.6	FUC	Cherbourg, France	50	North Atlantic Ocean
	FUE	Brest, France	50	
	FUO	Toulon, France	50	Mediterranean Sea
63.0	NHB	Kodiak, AK	50	
	NUD	Adak, AK	50	
63.85	FTA63	Paris, France	40	
64.2	NAM	Norfolk, VA	100	
	NPN	Guam	50	
	NSS	Washington, DC	200	
64.55	GBV20	Rugby, England	60	
64.9	CAD3A	Magallanes, Chile	20	London/Paris
	SOA90	Radom, Poland	40	
65.25	IRC	Rome, Italy	15	
65.8	FUE	Brest, France	15	
65.9	GXH	Thurso, Scotland	50	North Atlantic Ocean
	NST	Londonderry, N. Ireland	25	Kenitra
65.95	NAW	Guantanamo Bay	50	

kHz	Call	Location	Power (kW)	Remarks
	NPG	San Francisco, CA	50	
	NPL	San Diego, CA	50	
	NPO	Luzon, Philippines	500	Pacific Ocean
66.3	RDW	Moscow, USSR	30	
	RGA	Khabarovsk, USSR	30	
67.0	SLZ	Kristinehamn, Sweden	10	
67.5	AAH	Seattle, WA	10	
68.0	GBY20	Rugby, England	80	
	UPV	Ostrov Dickson, USSR	50	into Kara Sea
68.9	XPH	Thule, Greenland	25	Alert, Canada
69.1	RCK	Novosibirsk, USSR	30	
	UIK	Vladivostok, USSR	17.5	
69.7	DKQ	Koenigs Wusterhausen, G.D.R.	20	Sofia/Bucharest
69.8	IRD	Rome, Italy	15	
	NPC	Seattle, WA	50	
	NPG	San Franc.sco, CA	50	
72.0	EVA2	Baku, USSR	20	
72.1	FTA72	St. Assise, France	45	Lisbon
72.3	EVA2	Baku, USSR	20	
	RLQ	Mys Schmidt, USSR	30	
72.45	EAA	Aranjuez, Spain	8	Paris/London
72.5	ARL	Karachi, Pakistan	20	Arabian Sea
72.85	LCE	Jeloey, Norway	45	
72.9	ORL28	Ruiselede, Belgium	10	Oslo/Barcelona
73.25	NHB	Kodiak, AK	50	
	NPR	Dutch Harbor, AK	50	
	MTO21	Crimond, England	40	
73.6	CFH	Halifax, NS	250	North Atlantic Ocean
	VTH	Bombay, India	100	Arabian Sea
73.85	OEV22	Deutsch Altnbg., Austria	30	
74.2	GYD	London, England	10	North Sea
	MTO21	Crimond, England	40	
74.5	DKQ2	Koenigs Wusterhausen, G.D.R.	20	Sofia/Bucharest
74.55	SAY	Karlsborg, Sweden	10	
74.9	CAC3A	Talcahuano, Chile	20	Santiago
75.0	HBG	Geneveprangins, Switzerld.	40	
75.25	ESJ	Tallinn, USSR	10	
75.6	OXE21	Skamlebaek, Denmark	20	
75.95	GYW	Gibraltar	40	
	NDI	Naha, Okinawa	100	
	NDT4	Kami Seya, Japan	50	
	NPM	Pearl Harbor, HI	100	
	NPN	Guam	100	
	NPU	Pago Pago, Samoa	50	
76.35	TFK	Reykjavik, Iceland	50	North Atlantic Ocean
	SNA20	Radom, Poland	40	Vienna
76.75	RRW	Moscow, USSR	30	
77.15	NAM	Norfolk, VA	50	Atlantic Ocean
	NWP	Argentia, NFLD	50	Atlantic Ocean
77.5	DCF77	Mainflingen, G.F.R.	10	
	GLB20	Ongar, England	10	
77.82	GYA2	London, England	40	
	MTO21	Crimond, England	40	
77.85	HAU	Szekesfehervar, Hungary	30	
78.9	NPM	Pearl Harbor, HI	50	

kHz	Call	Location	Power (kW)	Remarks
	NPN	Guam	50	
79.5	FTA79	Lyon, France	5	
	FTA79	St. Assise, France	5	
79.8	RMT	Mys Chelyuskin, USSR	30	
79.95	VTO	Vizagapatam, India	50	Bay of Bengal
80.05	TFK	Reykjavik, Iceland	50	North Atlantic Ocean
80.3	GLO20	Ongar, England	15	
80.5	SNA30	Radom, Poland	40	
80.65	NGR	Kato Soli, Greece	100	
	NPC	Seattle, WA	50	
81.0	GYN	London, England	40	
81.35	SNA40	Radom, Poland	40	Stockholm/Oslo
81.7	FTA81	Paris, France	10	
	NBA	Balboa, Canal Zone	100	
	NPL	San Diego, CA	15	
82.05	HBA	Berne, Switzerland	40	
82.4	YTA	Belgrade, Yugoslavia	40	
82.75	GYB	London, England	40	
	GYW	Gibraltar	40	
83.1	OFA83	Nummela, Finland	40	
83.45	OLT8	Prague, Czechoslovakia	5	
83.45	AAH	Seattle, WA	50	
83.8	FTA83	Lyon, France	45	
84.0	JTA22	Ulan Bator, Mongolia	20	
86.0	NPO	Manila, Philippines	500	Guam/Pearl Harbor
86.05	TFK	Reykjavik, Iceland	50	Greenland/NFLD
86.6	SAW	Karlsborg, Sweden	40	
	RRX	Tchita, USSR	40	
86.95	GYW	Gibraltar	40	Atlantic Ocean
87.3	DEA	Bonn, G.F.R.	10	
	ALC	Fairbanks, AK	6	Nome/Anchorage
	ALC22	Nome, AK	15	Fairbanks
87.4	LCU	Jeloey, Norway	15	
87.65	DCF87	Mainflingen, G.F.R.	5	
	RNO	Moscow, USSR	30	
87.8	YQI	Constantza, Romania	20	
88.0	GZO	Hong Kong	40	
	GZV	Mauritius	20	Indian Ocean
	NSS	Washington, DC	50	
	NST	Londonderry, N. Ireland	25	
	OLT	Prague, Czechoslovakia	25	
88.7	PER	Kootwijk, Netherlands	10	
89.4	HBB2	Berne, Switzerland	40	
89.75	RNO	Moscow, USSR	30	
90.1	YTA4	Belgrade, Yugoslavia	40	
	ZRH	Fisantekraal, S. Africa	50	South Atlantic Ocean
91.15	FTA91	Lyon, France	150	
	JMC	Tokyo, Japan	10	
91.5	UIK	Vladivostok, USSR	17.5	
91.85	OLT9	Prague, Czechoslovakia	25	Moscow
92.0	FUC	Cherbourg, France	15	
	FUE	Brest, France	15	
92.2	OLT6	Prague, Czechoslovakia	25	
92.25	SPL	Gdynia, Poland	5	
93.0	FTA93	St. Assise, France	5	

kHz	Call	Location	Power (kW)	Remarks
93.6	IRE	Rome, Italy	15	Mediterranean Sea
93.9	FUO	Toulon, France	15	Mediterranean Sea
93.95	ALC	Fairbanks, AK	15	Nome/Anchorage
	ALC22	Nome, AK	6	Fairbanks
94.3	VTI	Bombay, India	100	Indian Ocean
94.5	DKQ3	Koenigs Wusterhausen, G.D.R.	10	Budapest/Moscow
94.65	LCT	Jeloey, Norway	15	
95.0	NPG	San Francisco, CA	50	
95.7	RAU	Tashkent, USSR	25	
96.05	HBB	Berne, Switzerland	40	
96.2	ARL	Karachi, Pakistan	20	Arabian Sea
97.1	DCF97	Mainflingen, G.F.R.	5	
97.45	GYW	Gibraltar	10	
97.7	OLT5	Prague, Czechoslovakia	25	Bucharest
97.8	RTO	Moscow, USSR	100	
98.0	UPV	Ostrov Dickson, USSR	50	Arctic Ocean
98.15	FTA98	Lyon, France	45	
	NPM	Pearl Harbor, HI	50	Pacific Ocean
98.5	TAB	Ankara, Turkey	15	
98.85	OLT2	Prague, Czechoslovakia	40	
99.2	FFK	St. Nazaire, France	10	
99.55	OEV33	Deutsch Altnbg., Austria	40	
99.7	DIU	Koenigs Wusterhausen, G.D.R.	10	
99.9	CAA2A	Santiago, Chile	20	

Appendix B
Suppliers

* Designates catalogue available by request
* Antennas LF/VLF
>   Tracor Inc.
>   Instruments Division
>   6500 Tracor Lane
>   Austin, Texas 78721
* Batteries (gel-cell, chargers, etc.)
>   Globe Union, Inc.
>   5757 North Green Bay Avenue
>   Milwaukee, Wisconsin 53201

   Calculators (single layer coil winding)
>   American Radio Relay League
>   Newington, Connecticut 06111
* Capacitors (tuning for the Basic SEA and SES Receiver)
>   Arco Electronics
>   Community Drive
>   Great Neck, New York 11022

   Chart Paper (recorder paper)
>   Graphic Controls Corporation
>   Recording Chart Division
>   189 Van Rensselaer Street
>   Buffalo, New York 14240

Single source of supply for all chart requirements (Ask for catalogue 26 or latest edition)

* Circuit Boards Full size etched circuit boards available for the Deluxe SES Receiver (Fig. 3-4). Complete with pure tin G-10 electroless backing

* Coax (semi-rigid) Wire Concepts
198 Passaic Avenue
Fairfield, N.J. 07006

undrilled available from the author at $15.00 p.p. Included also is a sizable piece of pure tin plate shielding that can be used for building a suitable enclosure as seen in Fig. 5-1. Send $15.00, check or money order, to:

Carl M. Chernan WA3UER
1135 Constitution Drive
Tarentum, Pa. 15084

Coils (For all receivers described in Chapter 3)
Available from: J. W. Miller Company
Distributor: G. R. Whitehouse & Company
11 Newbury Drive
Amherst, New Hampshire 03031
Coils (toroid type)-G. R. Whitehouse Company
Coils (for LF-VLF converter in Chapter VI)
Caddell Coil Corporation
35 Main Street
Poultney, Vermont 05764

* Converters
Janel Laboratories
3312 S.E. Van Buren Boulevard
Corvallis, Oregon 97330
(LF/VLF converter kit available in near future)

* Filters (active filters) Datel Systems, Inc.
1020G Turnpike St., Building S.
Canton, Massachusetts 02021

Filters Ceramic-(For the Deluxe SES Receiver in Chapter 3 (Fig. 3-3).

Vernitron Piezolectric Div.
232 Forbes Road
Bedford, Ohio 44146

General Parts Burstein-Applebee
              3199 Mercier
              Kansas City, Mo. 64111
* Instrument Amplifier (Model LA-1)
              Abbear Cal, Inc.
              123-17A Gray Avenue
              Santa Barbara, Calif. 93101
* Magnets (circular) Edmund Scientific Company
              Edscorp Building
              Barrington, New Jersey 08007
* Plastic Tubing United States Plastic Corp.
              1550 Elidia Road
              Lima, Ohio 45805
              (PVC tubing, connectors, etc.)
  Power Supplies (kit)-ADVA Electronics
              Box 4181F
              Woodside, California 94062
* Preamplifiers (LF/VLF) kit-International Crystal Mfg. Co.
              10 North Lee
              Oklahoma City, Oklahoma 73102
  Receivers (new-LF/VLF)
                   * National Radio Company
                   Washington Street
                   Melrose, Massachusetts 02176
                   *Tracor, Inc.
                   Instruments Division
                   6500 Tracor Lane
                   Austin, Texas 78721

                   * R. L. Drake Company
                   540 Richard Street
                   Miamisburg, Ohio 45342
  (Surplus LF/VLF)
                   Fair Radio Sales, Inc.
                   1016 East Eureka Street
                   Lima, Ohio 45802

                   Columbia Electronic Sales
                   P. O. Box 9266
                   North Hollywood, California 91609

Integrated Circuits Ferranti Electric, Inc.
    East Beth Page Road
    Plainview, N.Y. 11803
    (receiver IC)
Model 2750 (Simpson)
    Simpson Electric Company
    5200 W. Kinzie Street
    Chicago, Illinois 60610
(Ask for catalogue and local representative)
Recorders (used and new surplus; miniature strip chart and others)
    Edlie Electronics
    2700 Hempstead Turnpike
    Levittown, L. I., New York 11756

    Herbach & Rademan
    401 East Erie Avenue
    Philadelphia, Pa. 19134
Recorders (new) Miniature Strip Chart:
    Rustrak-Model 288
    Gulton Industries, Inc.
    Gulton Industrial Park
    East Greenwich, Rhode Island 02818
(Ask for Catalogue and local representative)
Transistors (Fairchild SE-4022, 2N5129)
    Poly Packs
    P.O. Box 942M
    Lynfield, Mass. 01940

Index